Fourier Analysis
of Numerical
Approximations of
Hyperbolic Equations

SIAM Studies in Applied Mathematics

JOHN A. NOHEL, Managing Editor

This series of monographs focuses on mathematics and its applications to problems of current concern to industry, government, and society. These monographs will be of interest to applied mathematicians, numerical analysts, statisticians, engineers, and scientists who have an active need to learn useful methodology for problem solving.

The first four titles in this series are: *Lie-Bäcklund Transformations in Applications*, by Robert L. Anderson and Nail H. Ibragimov; *Methods and Applications of Interval Analysis*, by Ramon E. Moore; *Ill-Posed Problems for Integrodifferential Equations in Mechanics and Electromagnetic Theory*, by Frederick Bloom; and *Solitons and the Inverse Scattering Transform*, by Mark J. Ablowitz and Harvey Segur.

*Robert Vichnevetsky
and John B. Bowles*

FOURIER ANALYSIS OF NUMERICAL APPROXIMATIONS OF HYPERBOLIC EQUATIONS

**with a Foreword by
Garrett Birkhoff**

siam *Philadelphia/1982*

Copyright © 1982 by Society for Industrial and Applied Mathematics. All rights reserved.

Library of Congress Catalog Card Number: 81-85699
ISBN: 0-89871-181-9

Contents

Foreword ix
Preface xi

Chapter 1. Introduction 1

  1.1 Introduction 1
  1.2 Linear models of hyperbolic equations 2
  1.3 Approximation 4
  1.4 Finite element Galerkin semi-discretizations 5
  1.5 Toeplitz operators 7
  1.6 B-spline Galerkin semi-discretizations 8
  1.7 Fully discrete approximations 11
  1.8 Fourier analysis 12
  1.9 $\mathscr{L}_2$-norms and Parseval's equality 12
  1.10 Relationship between continuous and discrete transforms . 13
  1.11 Fourier analysis of numerical approximations 16
  1.12 Sampling of the initial data 16

Chapter 2. Fourier Analysis of the Accuracy of Semi-Discretizations 19

  2.1 Introduction 19
  2.2 Sinusoidal trial solutions 19
  2.3 Fourier transforms and global errors 23
  2.4 Relation to classical truncation error analysis . . . 24
  2.5 Finite element Galerkin semi-discretizations 26
  2.6 A generalization 27
  2.7 Semi-discretization of a conservation law 29
  2.8 Time frequency 30
  2.9 The wave equation 31
  2.10 Implicit semi-discretizations of the wave equation . . 33

Chapter 3. Higher Order Semi-Discretizations 35

  3.1 Synthesis in the frequency domain 35
  3.2 Velocity error 37
  3.3 Limiting case ($K \to \infty$) 37
  3.4 Relation to the cardinal function 37

CONTENTS

3.5	B-spline Galerkin semi-discretizations	40
3.6	An equivalence of bases	40
3.7	Equivalence with collocation	41
3.8	Fourier analysis of the algorithms obtained with B-splines	43
3.9	Convergence rates	43
3.10	B-spline semi-discretizations of the wave equation	46
3.11	Analysis	48

Chapter 4. Full Discretizations — 51

4.1	Fourier analysis	51
4.2	Stability	54
4.3	Velocity and amplitude error	58
4.4	Examples	59
4.5	Time marching methods for second order equations	61

Chapter 5. Damping, Diffusion and Filtering — 63

5.1	Spurious diffusion	63
5.2	Limitations of the spurious diffusion model	65
5.3	Spurious diffusion in discretizations of the wave equation	66
5.4	Filtering	69
5.5	Low pass filters	70
5.6	Flat low pass filters	70
5.7	Fast Fourier transform filtering	71
5.8	Effect of filtering on damping and numerical stability	74

Chapter 6. Group Velocity — 75

6.1	Introduction	75
6.2	Group velocity and energy propagation	77
6.3	Group velocity of the simple 3-point finite differences semi-discretization	77
6.4	Group velocity of other 3-point semi-discretizations	78
6.5	Wave analysis	80
6.6	Wave analysis of other semi-discretizations	82

Chapter 7. Time-Fourier Transforms — 85

7.1	Introduction	85
7.2	Numerical phase velocity and wavelength	87
7.3	Relation to x-Fourier transforms	90
7.4	Cut-off frequency	92
7.5	Energy	95
7.6	Group velocity of the two fundamental types of solutions	95

7.7	Reflection at a downstream boundary	97
7.8	Convergence rates	99
7.9	A 3-point boundary formula	101

Chapter 8. Fourier Analysis and $\mathscr{L}_2$-Norm of the Global Error ... 103

8.1	Introduction	103
8.2	Examples	104
8.3	Relation with convergence rates analysis	107
8.4	Asymptotic approximations	107

Chapter 9. Spectral Methods ... 109

9.1	Introduction	109
9.2	Truncated Fourier series method	110
9.3	Analysis	111
9.4	Fourier "collocation" method	112

Chapter 10. Equations in Two Dimensions: Anisotropy ... 115

10.1	The advection equation in two dimensions	115
10.2	Anisotropy of the approximation on a square grid	116
10.3	Analysis of other explicit formulae	119
10.4	Implicit approximations	120
10.5	The wave equation in the plane	122
10.6	9-point semi-discretizations of the wave equation	128

Bibliography ... 131

Index ... 137

Foreword

For almost 200 years, Fourier analysis has been an indispensable tool for constructing analytical solutions of the partial differential equations of mathematical physics. Today it has become basic for analyzing their numerical solutions as well. This is evident from even a casual inspection of such classics as the books by Forsythe–Wasow, Varga, and Richtmyer–Morton.

But it is with hyperbolic equations that Fourier analysis finds its full power. The propagation of errors that occur in the process of numerical approximation may often be best described by concepts which originated in the study of sinusoidal waves by mathematical physicists.

Vichnevetsky and Bowles have brought together in the present book recent results in this general area which they and others have established in the past 15 years or so, but which were available so far only as journal articles or research reports. Bringing them together has underscored the existence of a common conceptual background which provides a basis for analyzing many questions in the approximation of hyperbolic equations.

In particular, the wave theory/group velocity analysis reveals the existence of two types of numerical solutions which propagate information in opposite directions, and provides the mathematics needed to describe and explain spurious reflection at boundaries. In a more practical vein, the use of stability charts in the interpretation of von Neumann's stability conditions, the use of numerical filtering, and investigations of anisotropy in two-dimensional computations suggest useful applications.

Although rigorously applicable only to difference and finite element approximations on a uniform mesh, and to linear differential equations with constant coefficients, the ideas which are suggested have a far wider range of relevance.

The style is lively but simple (neither Banach spaces nor hard theorems from classical analysis are treated). As a result, this book is a very readable yet somewhat unique text for students and workers in physics, applied mathematics, and computing who wish to acquire a general understanding of recent advances in error analysis.

GARRETT BIRKHOFF

Preface

There has been a growing interest, over the past decade or so, in the use of Fourier analysis to examine questions of accuracy and stability in numerical methods for hyperbolic equations. What had started at first as a set of individual, disconnected results published in the U.S. and Europe was becoming the identifiable single body of a more general theory. A great deal of material had appeared in the form of research papers and reports, and one of the original intents of this book was to bring together those results in a single place for easy reference. As is often the case, the whole turned out to be more than the sum of the parts: several new results have come to light during the assembly process, and have been suitably developed in the text.

A distinctive aspect of numerical methods for hyperbolic equations is that they introduce errors which distort the physical nature of the phenomena under study. Describing those errors by invoking concepts which originated in mathematical physics, such as energy propagation, dispersion and diffusion thus proves to be a most enlightening approach. In this respect, Fourier analysis provides an indispensable tool. This comes as no surprise: the analytical development of trigonometric series and integrals through the past two and a half centuries was often motivated by, and closely related to, the concurrent development of the partial differential equations of physics. In applying Fourier methods to the study of numerical discretizations of hyperbolic equations, one gets the feeling, by no coincidence, that the analysis is not an invention, but rather the rediscovery of a natural relationship that exists between the two.

This book should provide useful reference material to those who are engaged in one of the multiple aspects of computational fluid dynamics. It is intended for physicists and engineers who work with computers in the analysis of problems in such diverse fields as hydraulics, gas dynamics, plasma physics, numerical weather prediction and transport processes in chemical and civil engineering and who want or need to understand the implications of the approximations which they have used. It is also intended for applied mathematicians who are concerned with the more theoretical aspects of the computation.

Part of this text grew out of the notes of a course given by the senior author at Rutgers University to graduate students in computer science, physics and engineering, as a preparation for the analysis of numerical algorithms that they were using in their own research. Enough of the relevant aspects of discrete Fourier transforms have been incorporated to produce what comes close to a self-contained textbook for a one-semester course.

Needless to say, readers and students should be reasonably familiar with elements of the standard literature on numerical methods for hyperbolic equations. Exposure to at least one of the disciplines related to fluid dynamics or physics whence those equations originate is desirable. Finally, knowledge of the practical aspects of Fourier transforms, at the level at which they are taught to engineers (for example, Papoulis (1962)) is assumed.

Grateful acknowledgments are due to Professor Vassilios Dougalis who read a preliminary manuscript and made several useful suggestions that were incorporated in the final text.

R. VICHNEVETSKY
J. B. BOWLES

Chapter 1

Introduction

1.1 Introduction. The discovery of partial differential equations in the first half of the eighteenth century was preceded by discrete models of the vibrating string, described by John Bernoulli, in which a finite number of equidistant point masses were attached to a taut, but otherwise massless string. Passing to the limit was done by d'Alembert, who thus obtained what we now often call the one-dimensional wave equation:

(1.1) $$\frac{\partial^2 u}{\partial t^2} = c^2 \frac{\partial^2 u}{\partial x^2}.$$

In 1759, Lagrange used the same approach, solving the equations for a discrete model and then passing to the limit, to show that an analytic solution of that equation could be obtained as a trigonometric series.

A half century later, Fourier repeated the process, trigonometric series and all, to derive the heat equation

$$\frac{\partial u}{\partial t} = \sigma \frac{\partial^2 u}{\partial x^2}$$

and its solution, meanwhile leaving his name attached to both.

Up to that point, the calculations were aimed at deriving analytic results for continuous processes, and discrete models were used primarily as an intermediate tool toward that end. The concept of using discrete models as a means of obtaining *approximate solutions* of partial differential equations came later. One of its first mentions is found in Rayleigh's *Theory of Sound*. Table 1.1, extracted from that work, gives the ratio of the frequency of the first harmonic of a string of beads to the frequency of the corresponding continuous string. It was used by Rayleigh to illustrate the approximating power of finite differences, pointing out that "... we might by sufficiently multiplying the number of parts arrive at a system, still of finite freedom, but capable of any desired accuracy ...". In today's terminology, he would have said that "the approximation is convergent".

TABLE 1.1

m	1	2	3	4	9	19	39
$\dfrac{2(m+1)}{\pi}\sin\dfrac{\pi}{2(m+1)}$	.9003	.9549	.9745	.9836	.9959	.9990	.9997

This is perhaps one of the first uses of Fourier series to analyze the *accuracy* of the finite difference approximation of a partial differential equation. In this vein, the present text may be considered as a systematic use of Fourier analysis to investigate several questions related to errors in the numerical approximation of hyperbolic equations.

1.2. Linear models of hyperbolic equations. The simple advection equation

$$(1.2) \qquad \frac{\partial U}{\partial t} + c\frac{\partial U}{\partial x} = 0,$$

where c, a constant, has the dimensions of a velocity, is often used as a model for the description of numerical approximations of hyperbolic systems. How this comes about deserves some explanation.

The general form of a hyperbolic system of equations is

$$(1.3) \qquad \frac{\partial U}{\partial t} + F\frac{\partial U}{\partial x} = G,$$

where $U(x, t)$ is a vector of m functions of space x and time t, F is a square matrix of functions of U, and G is an m-vector of given forcing functions.

For a system to be hyperbolic, the matrix F must have m real eigenvalues and m distinct eigenvectors. Hyperbolic systems may be transformed to their characteristic form in the following manner: Let W_k be the kth left eigenvector of F,

$$(1.4) \qquad W_k^T F = W_k^T \nu_k,$$

where ν_k is the kth eigenvalue of F. Then, multiplying (1.3) to the left by W_k^T yields

$$W_k^T\left(\frac{\partial U}{\partial t} + F\frac{\partial U}{\partial x}\right) = W_k^T\left(\frac{\partial}{\partial t} + \nu_k\frac{\partial}{\partial x}\right)U = W_k^T G,$$

whence (1.3) becomes

$$(1.5) \qquad W_k^T \mathcal{D}_k U = W_k^T G,$$

where

$$(1.6) \qquad \mathcal{D}_k \equiv \frac{\partial}{\partial t} + \nu_k\frac{\partial}{\partial x} = \frac{d}{dt}\bigg|_k$$

INTRODUCTION

is a total derivative (also called a directional derivative) in the direction defined by

(1.7) $$\frac{dx}{dt} = \nu_k.$$

The lines defined by this direction field are *characteristic lines* of the equation, and to a hyperbolic system of order m there correspond m families of distinct characteristic lines.

When the system is linear with constant coefficients ($\boldsymbol{F}$ = a constant matrix), then we may define the new dependent variables

(1.8) $$V_k(x, t) = W_k^T U(x, t),$$

whereupon (1.5) becomes

(1.9) $$\frac{\partial V_k}{\partial t} + \nu_k \frac{\partial V_k}{\partial x} = W_k^T G \equiv H_k, \quad k = 1, 2, \cdots, m,$$

which is a system of m uncoupled, first order hyperbolic equations.

If (1.3) were to be approximated by some numerical algorithm on a fixed grid of points in (x, t), then the result would be identical to that obtained if the same algorithm were applied independently to the m equations (1.9), and the approximation to U subsequently recovered by the inverse transformation:

(1.10) $$U = \begin{bmatrix} W_1^T \\ \vdots \\ W_m^T \end{bmatrix}^{-1} \cdot \begin{bmatrix} V_1 \\ \vdots \\ V_m \end{bmatrix}.$$

Analyzing accuracy and stability of the numerical process may thus be conveniently carried out by separately analyzing the effect of the approximation upon the m independent equations (1.9). Moreover, since stability and accuracy are both concerned with errors, and since errors are perturbations, one may often ignore the right-hand side of (1.9) and consider

(1.11) $$\frac{\partial V_k}{\partial t} + \nu_k \frac{\partial V_k}{\partial x} = 0, \quad k = 1, 2, \cdots, m$$

as the model to be used in the analysis. This is precisely what (1.2) represents, with c taking on all the values $\nu_1, \nu_2, \cdots, \nu_m$.

When $\boldsymbol{F}$ is dependent upon U (i.e., the system is quasilinear), it is still (with caution) possible to use the same approach. The system (1.2) with constant c is then taken as a locally linearized model of the equation, and the ν_k are the corresponding (assumed locally constant) characteristic values.

While obviously incomplete, using (1.2) as a model of numerical approximations of hyperbolic systems nevertheless offers a means whereby many phenomena related to errors may be conveniently displayed and analyzed. We shall use this model in most of this text.

As for boundary conditions, we shall (except in Chapters 7 and 9), assume that the domain of the equation is the entire real axis

(1.12) $\qquad -\infty < x < \infty.$

The spirit of this assumption is identical to that used in the standard von Neumann method of analysis of numerical stability, in which *local* modes of instability are mathematically described as holding *everywhere*.

1.3. Approximation. We define a regular rectangular grid in (x, t)

(1.13)
$$x_n = nh, \quad n = \cdots, -1, 0, 1, 2, \cdots,$$
$$t^j = j\Delta t, \quad j = 0, 1, 2, \cdots,$$

on which numerical approximations of U are sought. These numerical values are denoted by lower case letters:

(1.14) $\qquad u_n^j \simeq U(x_n, t^j).$

As we discussed before, the model which we shall use to describe approximation schemes is the simple advection equation

(1.15) $\qquad \dfrac{\partial U}{\partial t} + c \dfrac{\partial U}{\partial x} = 0$

on the entire real axis

(1.16) $\qquad -\infty < x < \infty.$

A convenient and common way to derive a numerical approximation is to first define the time-continuous functions

(1.17) $\qquad u_n(t) \simeq U(x_n, t)$

and to write the system of ordinary differential equations

(1.18) $\qquad \dfrac{du_n}{dt} = \text{approximation of } \left[-c \dfrac{\partial U}{\partial x} \right]_n.$

For example, if we use the standard 3-point central finite differences approximation to $\partial U/\partial x$ we obtain

(1.19) $\qquad \dfrac{du_n}{dt} = -c \left(\dfrac{u_{n+1} - u_{n-1}}{2h} \right), \quad n = \cdots, -1, 0, 1, 2, \cdots.$

Such a system of difference-differential equations is called a *semi-discretization* of the partial differential equation (1.15). When a semi-discretization is integrated exactly in time, then the overall method of approximation is called a *method of lines*. By extension, the procedure which consists in first approximating a partial differential equation by a semi-discretization such as (1.19), and then numerically approximating the solution of this semi-discretization by using a discrete time-marching method (as shall be described in § 1.7 below) is also sometimes called a method of lines (see, e.g., Hicks and Wei (1967), Madsen and Sincovec (1973), Hyman (1979)). The name "method of lines" had been introduced in the literature by Russian authors in the 1940's (see Faddeeva (1949)).

Space shift operator. We introduce the space shift operator $\boldsymbol{E}\cdot$, defined by the relation

$$(1.20) \qquad u_{n+1} \equiv \boldsymbol{E} \cdot u_n, \quad \text{or} \quad u_{n+k} \equiv \boldsymbol{E}^{(k)} \cdot u_n$$

which is read as "the operator $\boldsymbol{E}\cdot$ to the kth power". Then, we may rewrite the preceding approximation in symbolic notation as

$$(1.21) \qquad \frac{du_n}{dt} = \boldsymbol{A} \cdot u_n,$$

where $\boldsymbol{A}\cdot$ is the operator

$$(1.22) \qquad \boldsymbol{A} \cdot \equiv -c\left(\frac{\boldsymbol{E}-\boldsymbol{E}^{-1}}{2h}\right)\cdot .$$

1.4. Finite element Galerkin semi-discretizations. A set of basis functions $\{\varphi_n(x)\}$ which satisfy the collocation conditions

$$(1.23) \qquad \varphi_n(x_m) = \begin{cases} 1 & \text{when } m = n, \\ 0 & \text{when } m \neq n \end{cases}$$

is prescribed and an approximate solution is defined everywhere as

$$(1.24) \qquad u(x, t) = \sum_n \varphi_n(x) u_n(t).$$

The equation residual is defined as

$$(1.25) \qquad \mathcal{R} = \frac{\partial u}{\partial t} + c \frac{\partial u}{\partial x} = \sum_n \varphi_n \frac{du_n}{dt} + \frac{d\varphi_n}{dx} u_n.$$

Galerkin's method consists in asking that $\mathcal{R}$ be orthogonal to the basis functions

$$(1.26) \quad \langle \varphi_n, \mathcal{R} \rangle \equiv \int \varphi_n(x) \mathcal{R}(x, t)\, dx = 0 \quad \text{for } n = \cdots, -1, 0, 1, 2, \cdots.$$

This is equivalent to choosing the $\{du_n/dt\}$ at all times so as to minimize

$$(1.27) \qquad \langle \mathcal{R}, \mathcal{R} \rangle \equiv \int \mathcal{R}^2\, dx.$$

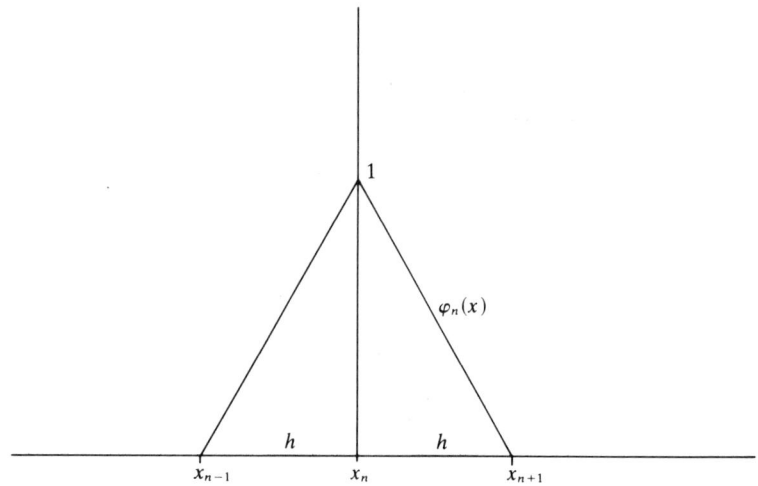

FIG. 1.1. *Linear basis functions.*

The simplest form is provided with linear elements; the basis functions are (see Fig. 1.1)

$$(1.28) \qquad \varphi_n(x) = \begin{cases} 1 - \left|\dfrac{x - x_n}{h}\right| & \text{for } |x - x_n| \leq h, \\ 0 & \text{elsewhere.} \end{cases}$$

The application of (1.26) then results, after evaluation of the coefficients, in

$$(1.29) \qquad \frac{h}{6}\left(\frac{du_{n-1}}{dt} + 4\frac{du_n}{dt} + \frac{du_{n+1}}{dt}\right) = -c\left(\frac{u_{n+1} - u_{n-1}}{2}\right),$$

or, dividing by h,

$$(1.30) \qquad \frac{1}{6}\left(\frac{du_{n-1}}{dt} + 4\frac{du_n}{dt} + \frac{du_{n+1}}{dt}\right) = -c\left(\frac{u_{n+1} - u_{n-1}}{2h}\right).$$

Note that the right-hand side is identical to that obtained with 3-point central finite differences. Note also that the left-hand side of (1.30) is equal to a mean value of $\partial u/\partial t$ in $(x_n - h, x_n + h)$ obtained by the application of Simpson's rule of integration.

This semi-discretization can be rewritten in operator form as

$$(1.31) \qquad \mathbf{A}_1 \cdot \frac{du_n}{dt} = \mathbf{A}_2 \cdot u_n,$$

where $\boldsymbol{A}_1 \cdot$ and $\boldsymbol{A}_2 \cdot$ are the operators

(1.32a) $$\boldsymbol{A}_1 \cdot \equiv \tfrac{1}{6}(\boldsymbol{E}^{-1} + 4 + \boldsymbol{E}) \cdot = \sum_k a_{1,k} \boldsymbol{E}^k \cdot,$$

and

(1.32b) $$\boldsymbol{A}_2 \cdot \equiv -c\left(\frac{\boldsymbol{E} - \boldsymbol{E}^{-1}}{2h}\right) \cdot = \sum_k a_{2,k} \boldsymbol{E}^k \cdot.$$

A semi-discretization in which $\boldsymbol{A}_1$ is not a diagonal operator is called an *implicit* semi-discretization. The reason for this definition is that when an explicit time marching method is applied to such a semi-discretization, one obtains a system of implicit algebraic equations to be solved at each time step. (This will be shown later.)

We can still use the notation (1.21) to describe implicit semi-discretizations, by agreeing that, in this case, the operator $\boldsymbol{A} \cdot$ formally stands for $\boldsymbol{A}_1^{-1} \cdot \boldsymbol{A}_2 \cdot$.

We note that in (1.32)

(1.33) $$\sum_k a_{1,k} = 1.$$

It shall always be assumed (when it matters) that this normalizing relation holds in implicit semi-discretizations.

1.5. Toeplitz operators. A matrix whose elements are equal on any line parallel to the main diagonal is called a Toeplitz matrix. Thus, operators of the kind defined by (1.32), with coefficients $\{a_k\}$ independent of n are infinite Toeplitz matrices (also called Toeplitz operators). A property that we shall invoke later on is that *any two Toeplitz operators $T_1 \cdot$ and $T_2 \cdot$ commute*; i.e.,

(1.34) $$\boldsymbol{T}_1 \cdot \boldsymbol{T}_2 \cdot = \boldsymbol{T}_2 \cdot \boldsymbol{T}_1 \cdot.$$

Indeed, the necessary and sufficient condition for two operators (matrices) to commute is that they share the same eigenvectors (see, e.g., Noble (1969, p. 342)). It may be observed that, for real ω,

(1.35) $$\{e^{i\omega nh}\}$$

are the eigenvectors of any Toeplitz operator (see also § 2.2). They are thus the same for $\boldsymbol{T}_1 \cdot$ and $\boldsymbol{T}_2 \cdot$, which therefore commute. A simpler, somewhat trivial proof consists in constructing $\boldsymbol{T}_1 \cdot \boldsymbol{T}_2 \cdot$ and $\boldsymbol{T}_2 \cdot \boldsymbol{T}_1 \cdot$ and in verifying that their elements are identical.

The infinite vectors (1.35) also form the basis of functions which is implicit in Fourier transformation. Since operators become scalars when they operate on their own eigenvectors, Toeplitz operators have images in the Fourier domain that are scalar functions. The algebra of functions is of course more convenient than the algebra of operators, and therein lies one of the reasons for the power of Fourier analysis.

1.6. B-spline Galerkin semi-discretizations. Trial solutions $u(x, t)$ to be used with Galerkin's method may also be expressed as polynomial splines. (A polynomial spline of degree q is an interpolant between the "knots" $\{x_n, u_n\}$ that is a polynomial of degree q between those knots, and that has $(q-1)$ continuous derivatives across them.)

A convenient tool for the expression of u as a polynomial spline on a regular grid $\{x_n = nh\}$ is provided by a special set of basis functions, called B-splines, that were invented by Schoenberg (1946). They are defined as follows:

Consider the elementary (characteristic) function

(1.36) $$M_1(y) = \begin{cases} 1 & \text{for } |y| \leq \tfrac{1}{2}, \\ 0 & \text{for } |y| > \tfrac{1}{2}, \end{cases}$$

and its successive convolutions with itself

(1.37)
$$M_2(y) \equiv M_1(y) \otimes M_1(y) = \int M_1(y-\eta) M_1(\eta) \, d\eta,$$
$$\vdots$$
$$M_\mu(y) \equiv M_{\mu-1}(y) \otimes M_1(y) = \int M_{\mu-1}(y-\eta) M_1(\eta) \, d\eta.$$

These functions are shown in Fig. 1.2. It is easily verified that:
- $M_\mu(y)$ is a spline of degree $(\mu - 1)$,
- with knots located at the mesh points when μ is even, and midway between mesh points when μ is odd,
- with compact support:

(1.38) $$M_\mu(y) = 0 \quad \text{for } |y| > \frac{\mu}{2}.$$

On a regular mesh $\{x_n = nh\}$, B-spline basis functions are defined as

(1.39) $$\varphi_n^q(x) = M_{q+1}\left(\frac{x - x_n}{h}\right).$$

It may then be verified that

(1.40) $$u(x, t) \equiv \sum_n \varphi_n^q(x) v_n(t)$$

is a spline of degree q, with knots located at the mesh points when q is odd and midway between mesh points when q is even.

INTRODUCTION

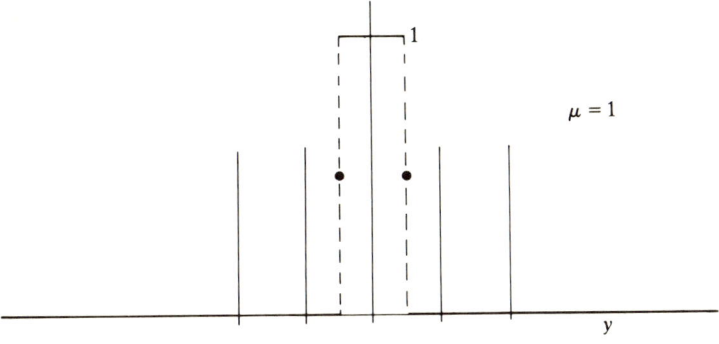

$\mu = 1$

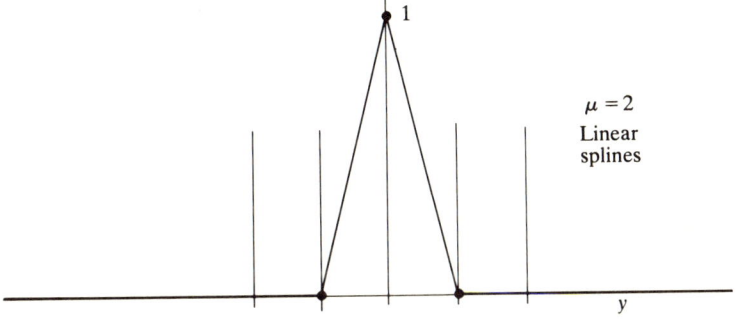

$\mu = 2$
Linear splines

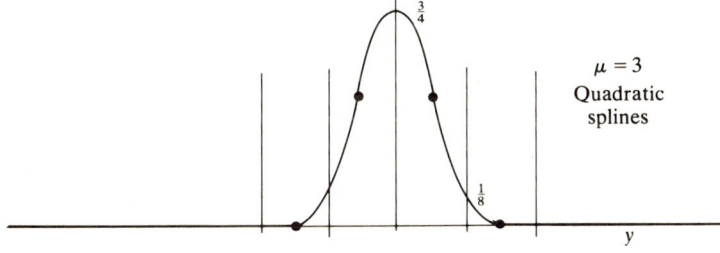

$\mu = 3$
Quadratic splines

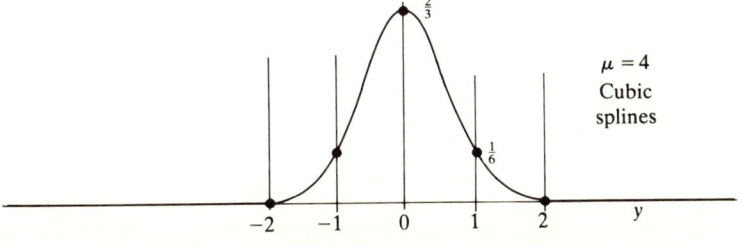

$\mu = 4$
Cubic splines

FIG. 1.2

Collocation at the mesh points is expressed by

(1.41)
$$u_n(t) = \sum_m \varphi_m^q(x_n) v_m$$

for every n, or, in operator notation,

(1.41')
$$u_n = \mathbf{T} \cdot v_n,$$

where $\mathbf{T} \cdot$ is the Toeplitz operator

$$\mathbf{T} \cdot \equiv \sum_{k=-K}^{K} \varphi_0^q(kh) \mathbf{E}^k \cdot$$

with

$$K = \begin{cases} \dfrac{q}{2} & \text{for } q \text{ even,} \\ \dfrac{q-1}{2} & \text{for } q \text{ odd.} \end{cases}$$

A B-spline Galerkin approximation of the solution of $U_t + cU_x = 0$ is obtained by forming the equation residual for the trial solution (1.40) and then expressing orthogonality with respect to the basis functions $\{\varphi_n^q(x)\}$. The result may be expressed as

(1.42)
$$\mathbf{A}_1 \cdot \frac{dv_n}{dt} = \mathbf{A}_2 \cdot v_n,$$

where the elements of $\mathbf{A}_1 \cdot$ are

$$a_{1,k} = \frac{1}{h} \langle \varphi_0^q(x), \varphi_k^q(x) \rangle = \frac{1}{h} \langle \varphi_0^q(x), \varphi_0^q(x+kh) \rangle$$

and those of $\mathbf{A}_2 \cdot$ are

$$a_{2,k} = -\frac{c}{h} \left\langle \varphi_0^q(x), \frac{d\varphi_k^q(x)}{dx} \right\rangle = \frac{c}{h} \left\langle \varphi_0^q(x), \frac{d\varphi_0^q(x+kh)}{dx} \right\rangle$$

(division by h is introduced to normalize $\mathbf{A}_1 \cdot$ per (1.33)). Those operators are banded:

$$a_{i,k} = 0 \quad \text{for } |k| > q.$$

The case $q = 1$ is identical to the linear finite element case (1.30), and the case $q = 2$ (quadratic splines with knots located midway between mesh points) results in:

(1.43)
$$\frac{1}{120}\frac{d}{dt}(v_{n-2}+26v_{n-1}+66v_n+26v_{n+1}+v_{n+2})$$
$$=-\frac{c}{24h}(v_{n+2}+10v_{n+1}-10v_{n-1}-v_{n-2}).$$

Note that, in contrast with finite difference and finite element methods, the coefficients $\{v_n(t)\}$ are not the nodal values $\{u_n(t)\}$, which may be inconvenient in practice. One may, however, derive an interesting theoretical equivalence between (1.42) which describes the evolution of the spline coefficients $\{v_n\}$ and an *identical* equation which describes the evolution of the nodal values $\{u_n\}$. This is developed in § 3.6 below.

1.7. Fully discrete approximations. To obtain a fully discrete numerical approximation, we apply to the semi-discrete system (1.21) a time-marching method for ordinary differential equations. For example, Euler's method is

(1.44) $$\mathbf{u}^{j+1} \equiv \{u_n^{j+1}\} = \mathbf{u}^j + \Delta t A \cdot \mathbf{u}^j.$$

When the semi-discretization is implicit, then (1.44) is actually

$$\mathbf{A}_1 \cdot \mathbf{u}^{j+1} = (\mathbf{A}_1 + \Delta t \mathbf{A}_2) \cdot \mathbf{u}^j,$$

which is a system of implicit linear equations with matrix $\mathbf{A}_1 \cdot$, to be solved for the discrete set of values $\mathbf{u}^{j+1} \equiv \{u_n^{j+1}, n = \cdots 0, 1, 2, \cdots\}$ at each time step.

Euler's *implicit* method is

(1.45) $$\mathbf{u}^{j+1} = \mathbf{u}^j + \Delta t A \cdot \mathbf{u}^{j+1} = (I + \Delta t A \cdot Z \cdot) \cdot \mathbf{u}^j,$$

where $Z \cdot$ is the time shift operator, formally defined by the relation

(1.46) $$\mathbf{u}^{j+1} \equiv Z \cdot \mathbf{u}^j.$$

This is solved at each time step as the system of equations

(1.47) $$(\mathbf{A}_1 - \Delta t \mathbf{A}_2) \cdot \mathbf{u}^{j+1} = \mathbf{A}_1 \cdot \mathbf{u}^j.$$

All time marching methods applied to (1.21) may be described in operator notation as

(1.48) $$\mathbf{u}^{j+1} = M(\Delta t A \cdot Z) \cdot \mathbf{u}^j,$$

where M is a polynomial in the operators Z, Z^{-1} and $\Delta t A \cdot$. For example, in the preceding examples M is given by

(1.49)   Euler explicit method: $M = I + \Delta t A \cdot$,

(1.50)   Euler implicit method: $M = I + \Delta t A \cdot Z \cdot$, and for the *leapfrog method*

(1.51) $$\mathbf{u}^{j+1} = \mathbf{u}^{j-1} + 2\Delta t \mathbf{A} \cdot \mathbf{u}^j$$

we have

(1.52) $$M = Z^{-1} + 2\Delta t \mathbf{A} \cdot .$$

1.8. Fourier analysis. The *Fourier transform* $\hat{U}(\omega, t)$ of $U(x, t)$ is, by definition, for a fixed t,

(1.53) $$\hat{U}(\omega, t) = \int_{-\infty}^{\infty} U(x, t) e^{-i\omega x} \, dx \equiv \mathscr{F}(U).$$

For this transform to exist, $U(x, t)$ must be in $\mathscr{L}_2$, i.e., it must be square summable

(1.54) $$\int_{-\infty}^{\infty} U^2(x, t) \, dx < \infty.$$

The inverse transformation is

(1.55) $$U(x, t) = \int_{-\infty}^{\infty} \hat{U}(\omega, t) e^{i\omega x} \frac{d\omega}{2\pi} \equiv \mathscr{F}^{-1}(\hat{U}).$$

The analogue of these relations for functions defined at discrete points only (such as $\{u_k(t)\}$ and $\{u_n^j\}$) is given by the *discrete Fourier transforms* defined as[1]

(1.56) $$\tilde{u}(\omega, t) = h \sum_{n=-\infty}^{\infty} u_n(t) e^{-i\omega x_n}$$

or

(1.57) $$\tilde{u}^j(\omega) = h \sum_{n=-\infty}^{\infty} u_n^j e^{-i\omega x_n}.$$

The inverse is

(1.58) $$u_n^j = \int_{-\pi/k}^{\pi/k} \tilde{u}^j(\omega) e^{i\omega x_n} \frac{d\omega}{2\pi}.$$

1.9. $\mathscr{L}_2$-norms and Parseval's equality. The $\mathscr{L}_2$-norm $\|U(x)\|_2$ of a function $U(x)$ is defined as

(1.59) $$\|U(x)\|_2 \equiv (\langle U(x), U(x) \rangle)^{1/2} = \left(\int_{-\infty}^{\infty} |U(x)|^2 \, dx \right)^{1/2}.$$

---

[1] The notation "square hat" ~ to define a *discrete* Fourier transform is ours. Standard texts use ^ for both continuous and discrete Fourier transforms. We find that using different symbols to distinguish between those two entities, which are really different, adds to the clarity.

The $\mathcal{L}_2$-norm of a function is related to its Fourier transform via Parseval's equality

(1.60) $$\|U(x)\|_2^2 \equiv \int_{-\infty}^{\infty} |U(x)|^2 \, dx = \int_{-\infty}^{\infty} |\hat{U}(\omega)|^2 \frac{d\omega}{2\pi} \equiv \|\hat{U}(\omega)\|_2^2.$$

In the context of an analysis of discrete approximations we shall also be interested in *discrete* $\mathcal{L}_2$-norms,

(1.61) $$\|u_n\|_2 \equiv \left(h \sum_{n=-\infty}^{\infty} |u_n|^2 \right)^{1/2}.$$

The relationship of these norms to the discrete Fourier transform of $\{u_n\}$ is the discrete form of Parseval's equality,

(1.62) $$\|u_n\|_2^2 \equiv h \sum_{n=-\infty}^{\infty} |u_n|^2 = \int_{-\pi/h}^{\pi/h} |\tilde{u}(\omega)|^2 \frac{d\omega}{2\pi} \equiv \|\tilde{u}(\omega)\|_2^2.$$

A proof of this relation is given in the following section.

1.10. Relationship between continuous and discrete transforms. Consider the discrete set $\{u_n\}$ and the continuous function derived from it by the relation

(1.63) $$u^*(x) = \sum_n \frac{\sin[\pi(x-x_n)/h]}{[\pi(x-x_n)/h]} u_n.$$

We may define the set of basis functions

(1.64) $$\Psi_n(x) = \frac{\sin[\pi(x-x_n)/h]}{[\pi(x-x_n)/h]}$$

and rewrite (1.63) as

(1.65) $$u^*(x) = \sum_n \Psi_n(x) u_n.$$

Since

(1.66) $$\Psi_n(x_m) = \begin{cases} 1 & \text{when } m = n, \\ 0 & \text{when } m \neq n, \end{cases}$$

it follows that

(1.67) $$u^*(x_n) = u_n.$$

Hence $u^*(x)$ is an interpolant between the discrete points $\{x_n; u_n\}$. The function $\Psi_0(x)$ is sometimes called the *cardinal function* (or Whittaker's cardinal

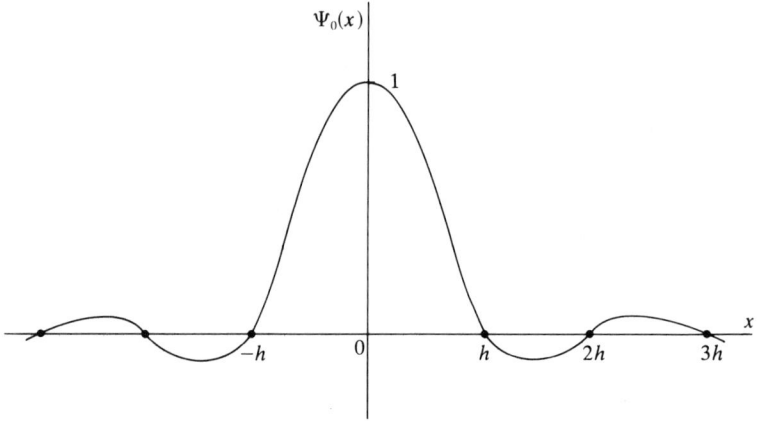

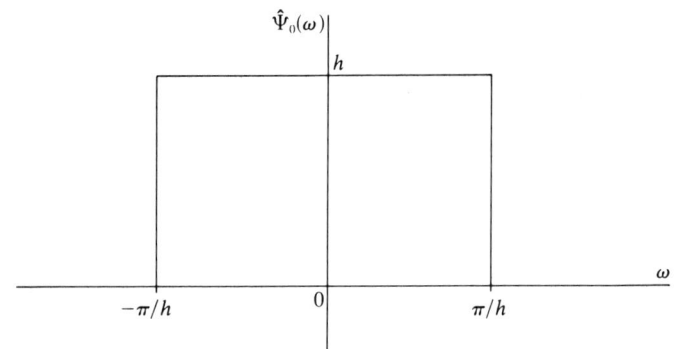

FIG. 1.3. *Whittaker's cardinal function and its Fourier transform.*

function, after its author (see Whittaker (1915)), and is shown in Fig. 1.3. It has the interesting Fourier transform:

$$\hat{\Psi}_0(\omega) = \begin{cases} h & \text{in } -\frac{\pi}{h} \leq \omega \leq \frac{\pi}{h}, \\ 0 & \text{elsewhere.} \end{cases} \tag{1.68}$$

Since $\Psi_n(x)$ is obtained by shifting $\Psi_0(x)$ by nh, we also have

$$\hat{\Psi}_n(\omega) = e^{-i\omega nh} \hat{\Psi}_0(\omega), \tag{1.69}$$

which results from a direct application of the shifting rule for Fourier transforms.

An important property of Whittaker's cardinal functions is that they form an orthogonal set; we find by application of the convolution theorem[2]

(1.70)
$$\langle \Psi_m, \Psi_n \rangle \equiv \int_{-\infty}^{\infty} \Psi_m(x) \Psi_n(x)\, dx$$
$$= \int_{-\infty}^{\infty} \hat{\Psi}_m(\omega) \hat{\Psi}_n(-\omega) \frac{d\omega}{2\pi}$$
$$= \int_{-\pi/h}^{\pi/h} h^2 e^{i\omega h(n-m)} \frac{d\omega}{2\pi}$$
$$= h \frac{\sin[(n-m)\pi]}{(n-m)\pi} = \begin{cases} h & \text{for } n = m, \\ 0 & \text{for } n \neq m. \end{cases}$$

We now return to the function $u^*(x)$. Its Fourier transform is easily found:

(1.71)
$$\hat{u}^*(\omega) = \hat{\Psi}_0(\omega) \sum_n u_n e^{-i\omega nh};$$

that is

(1.72)
$$\hat{u}^*(\omega) = \begin{cases} \bar{u}(\omega) & \text{in } -\frac{\pi}{h} < \omega < \frac{\pi}{h}, \\ 0 & \text{elsewhere.} \end{cases}$$

If we apply Parseval's equality (1.60) to the continuous function $u^*(x)$ we find

(1.73)
$$\|u^*(x)\|_2^2 = \int_{-\infty}^{\infty} |\hat{u}^*(\omega)|^2 \frac{d\omega}{2\pi} = \int_{-\pi/h}^{\pi/h} |\bar{u}(\omega)|^2 \frac{d\omega}{2\pi}.$$

But, also, *because of the orthogonality property* (1.70),

(1.74)
$$\|u^*(x)\|_2^2 = \sum_n u_n^2 \langle \Psi_n, \Psi_n \rangle + \sum_{n \neq m} u_n u_m \langle \Psi_n, \Psi_m \rangle$$
$$= h \sum_n u_n^2.$$

Thus, (1.73) is also equal to the square of the discrete $\mathcal{L}_2$-norm of $\{u_n\}$, as stated in (1.62).

This is an important result. The relations (1.71)–(1.73) which apply to the *continuous function* $u^*(x)$ are identical to the relations (1.56)–(1.62) which apply to the discrete set $\{u_n\}$. It is thus sometimes useful to consider a discrete set $\{u_n\}$ as the samples, taken h apart, of the continuous, band-limited function $u^*(x)$. Both $\{u_n\}$ and $u^*(x)$ carry the same information, and are analytically equivalent entities.

---

[2] See, e.g., Papoulis (1962, pp. 25–29).

The converse goes as follows:

A *band-limited function* is defined as a function $U(x)$ whose Fourier transform $\hat{U}(\omega)$ vanishes outside a finite band, say $|\omega| > B$. If this function is sampled:

$$\{u_n\} \equiv U(nh)$$

and reconstructed:

$$u^*(x) = \sum_n \Psi_n(x) u_n,$$

then $u^*(x)$ and $U(x)$ are identical if $h \leq \pi/B$. This result was proven by Whittaker in 1915, and later used by Shannon (1949) in an important contribution to communications theory.

1.11. Fourier analysis of numerical approximations. If $\hat{U}(\omega, t)$ is the Fourier transform of a solution of the model equation $U_t + cU_x = 0$, then we find by insertion in that equation that

(1.75) $$\frac{\partial \hat{U}}{\partial t} + ci\omega \hat{U} = 0.$$

This gives by analytic integration

(1.76) $$\hat{U}(\omega, t) = \hat{U}(\omega, 0) e^{-ic\omega t}$$

where $\hat{U}(\omega, 0)$ is the Fourier transform of the initial function:

(1.77) $$\hat{U}(\omega, 0) = \int_{-\infty}^{\infty} U(x, 0) e^{-i\omega x} dx.$$

The numerical solution may be described by its *discrete* Fourier transform $\bar{u}(\omega, t)$. Analyzing differences between $\bar{u}(\omega, t)$ and $\hat{U}(\omega, t)$ forms the basis for a Fourier analysis of the approximation.

We may identify two types of errors.

First, replacing the initial function $U(x, 0)$ by a discrete set $\{u_n(0)\}$ results in a loss of data which affects the high frequency content of the solution.

Second, the time evolution of the Fourier transforms $\bar{u}(\omega, t)$ and $\hat{U}(\omega, t)$ are different. It is mostly this difference which will form the basis for the detailed Fourier analysis which is presented in subsequent chapters.

1.12. Sampling of the initial data. The error introduced by the replacement of the initial function $U(x, 0)$ by the discrete set $\{u_n(0)\}$ may be analyzed in the frequency domain.

The usual initialization procedure consists in *sampling* the initial function, simply letting

(1.78) $$u_n(0) = U(nh, 0), \quad n = \cdots, -1, 0, 1, 2, \cdots.$$

When a continuous function is sampled on a regular grid $\{x_n = nh\}$ as in (1.78), the discrete Fourier transform of the discrete set $\{u_n\}$ is related to the Fourier transform of the original function U by the relation, akin to Poisson's sum formula (see, e.g. Papoulis (1962, p. 48))

$$(1.79) \qquad \bar{u}(\omega, 0) = \sum_{k=-\infty}^{\infty} \hat{U}(\omega + k\omega_0, 0)$$

where ω takes values in

$$(1.80) \qquad -\frac{\pi}{h} \leq \omega \leq \frac{\pi}{h}$$

and where

$$(1.81) \qquad \omega_0 = \frac{2\pi}{h}$$

is called the "sampling frequency".

If $\hat{U}(\omega, 0) = 0$ for $|\omega| > \pi/h$, i.e., if the initial function $U(x, 0)$ is suitably band-limited, then we have simply

$$(1.82) \qquad \bar{u}(\omega, 0) = \hat{U}(\omega, 0)$$

and there is no error introduced by the initial sampling.

But if $U(x, 0)$ is not so band-limited, then higher frequency components in $\hat{U}(\omega, 0)$ are folded by (1.79) into the band $[-\pi/h, \pi/h]$ of $\bar{u}(\omega, 0)$. This is called *aliasing*. (After sampling, high frequency components become indiscernible from their low frequency "alias" in $|\omega| \leq \pi/h$.)

Rather than using for $\{u_n(0)\}$ the sampled values of $U(x, 0)$, it is sometimes recommended that the high frequencies be eliminated altogether.[3] This is done by choosing

$$(1.83) \qquad \bar{u}(\omega, 0) = \hat{U}(\omega, 0) \quad \text{for } |\omega| \leq \frac{\pi}{h}$$

and subsequently computing the values of $\{u_n(0)\}$ by the inverse transform

$$(1.84) \qquad u_n(0) = \int_{-\pi/h}^{\pi/h} \bar{u}(\omega, 0) e^{i\omega x_n} \frac{d\omega}{2\pi}.$$

If more accurate, this is, however, a somewhat more complicated procedure since it requires that the Fourier transform $\hat{U}(\omega, 0)$ be obtained, either analytically or by some numerical approximation which does not introduce aliasing (or at least much less aliasing, for example, by using a much smaller value for h).

---

[3] See, e.g., Kreiss and Oliger (1973).

Chapter 2

Fourier Analysis of the Accuracy of Semi-Discretizations

2.1. Introduction.[1] We first analyze the error which is inherent to semi-discretizations, i.e., the error due to the approximation of the model equation

$$(2.1) \qquad \frac{\partial U}{\partial t} + c\frac{\partial U}{\partial x} = 0$$

by the system of difference-differential equations (or semi-discretizations)

$$(2.2) \qquad \frac{du_n}{dt} = \boldsymbol{A} \cdot u_n.$$

Recall from the notation in Chapter 1 that for explicit semi-discretizations the discrete operator $\boldsymbol{A} \cdot$ is of the form

$$\boldsymbol{A} \cdot \equiv \sum_k a_k \boldsymbol{E}^k \cdot .$$

For implicit semi-discretizations we have

$$(2.3) \qquad \boldsymbol{A} \cdot = \boldsymbol{A}_1^{-1} \cdot \boldsymbol{A}_2 \cdot ,$$

where $\boldsymbol{A}_1 \cdot$ and $\boldsymbol{A}_2 \cdot$ are both explicit, discrete operators, so that (2.2) is then equivalent to

$$(2.4) \qquad \boldsymbol{A}_1 \cdot \frac{du_n}{dt} = \boldsymbol{A}_2 \cdot u_n.$$

2.2. Sinusoidal trial solutions. The sinusoidal function

$$(2.5) \qquad u_{\omega,n}(t) = v_\omega(t)\, e^{i\omega x_n}$$

---

[1] Table 2.1 in § 2.7 summarizes the results of the next several sections.

is a solution of the semi-discretization (2.2) provided

(2.6) $$\frac{dv_\omega}{dt} = \hat{A}(\omega)v_\omega,$$

where $\hat{A}(\omega)$ is the function of ω defined by

(2.6′) $$\hat{A}(\omega) = \frac{\mathbf{A}_2 \cdot e^{i\omega x_n}}{\mathbf{A}_1 \cdot e^{i\omega x_n}}.$$

This function is called the *spectral function* of the operator $\mathbf{A}\cdot$ (the names *spectrum* and *symbol*[2] are also sometimes used to denote $\hat{A}(\omega)$). We have noted before that $\{e^{i\omega x_n}\} = \{e^{i\omega nh}\}$ are eigenvectors of any discrete (Toeplitz) operator $\mathbf{A}\cdot$: the spectral functions $\hat{A}(\omega)$ are precisely the corresponding eigenvalues.

We find by analytic integration of (2.6) that

(2.7) $$\begin{aligned} u_{\omega,n}(t) &= v_\omega(0)\, e^{\hat{A}(\omega)t}\, e^{i\omega x_n} \\ &= v_\omega(0)\, e^{\mathrm{Re}\hat{A}(\omega)t}\, e^{i(\omega x_n + \mathrm{Im}\hat{A}(\omega)t)}. \end{aligned}$$

Likewise, sinusoidal solutions of the exact equation (2.1) are

(2.8) $$U_\omega(x,t) = V_\omega(0)\, e^{i\omega(x-ct)}.$$

We thus find that sinusoidal functions, with appropriate time-dependent coefficients, are solutions of both the exact equation and its semi-discretizations. Such functions considered one ω at a time, are called "sinusoidal trial solutions", and the difference between the two provides a description of the discrepancy introduced by the semi-discrete approximation. (Their generalization into complete Fourier transforms is considered in § 2.3 and Chapter 8.) The expressions (2.7) and (2.8) differ in two respects.

(i) *Amplitude.* Whereas the amplitude of exact sinusoidal solutions remains constant with time,

(2.9) $$|V_\omega(t)| = |V_\omega(0)|,$$

that of numerical sinusoidal solutions does not necessarily remain constant. We have

(2.10) $$|v_\omega(t)| = |v_\omega(0)|\, e^{\mathrm{Re}\hat{A}(\omega)t}.$$

The condition

$$\mathrm{Re}\,\hat{A}(\omega) = 0 \quad \text{for all } \omega$$

is easily achieved (it suffices for $\mathbf{A}\cdot$ to be antisymmetric). Then the semi-discretization is called *conservative*, and

$$|v_\omega(t)| = |v_\omega(0)| \quad \text{for all } \omega.$$

---

[2] See, e.g., Thomee (1969).

When Re $\hat{A}(\omega) \leq 0$ for all ω, and the strict inequality holds for at least some ω, there is amplitude decay of the corresponding sinusoidal solution as

$$e^{\operatorname{Re}\hat{A}(\omega)t} < 1.$$

The semi-discretization is then called *dissipative*. When Re $\hat{A}(\omega) > 0$ for some ω, the amplitude of the corresponding sinusoidal solutions grows exponentially with time and the semi-discretization is *unstable*.

(ii) *Phase velocity error*. For the class of numerical methods which we have called "conservative", (Re $\hat{A}(\omega) \equiv 0$), the only discrepancy between numerical sinusoidal solutions and their exact counterparts is in the velocity at which they propagate. If we define the function

(2.11) $$c^*(\omega) \equiv \frac{-\operatorname{Im}\hat{A}(\omega)}{\omega}$$

then we may rewrite (2.7) as

(2.12) $$u_{\omega,n}(t) = v_\omega(0)\, e^{i\omega(x_n - c^*(\omega)t)}.$$

This shows that $c^*(\omega)$ is the velocity of propagation of numerical sinusoidal solutions. $c^*(\omega)$ is called the *phase velocity* in classical wave theory. We observe that *in contrast with exact solutions, c^* is dependent upon the frequency ω* (or the wavelength $\lambda = 2\pi/\omega$). This is illustrated by the following examples.

Example 2.1. For the simple explicit (3-point central finite differences, see (1.19)) semi-discretization of (2.1)

(2.13) $$\frac{du_n}{dt} = -c\left(\frac{u_{n+1} - u_{n-1}}{2h}\right),$$

we find

$$\hat{A}(\omega) = -ic \cdot \frac{\sin(\omega h)}{h}.$$

Since

$$\operatorname{Re}\hat{A}(\omega) \equiv 0,$$

the approximation is conservative. The phase velocity is found by (2.11):

(2.13') $$c^*(\omega) = c\left(\frac{\sin(\omega h)}{\omega h}\right).$$

This function is illustrated in Figs. 2.1a and 2.1b and in Table 2.1, p. 30.

Example 2.2. The *implicit* 2-point (or "box") semi-discretization

(2.14) $$\frac{1}{2}\left(\frac{du_n}{dt} + \frac{du_{n+1}}{dt}\right) = -c\left(\frac{u_{n+1} - u_n}{h}\right)$$

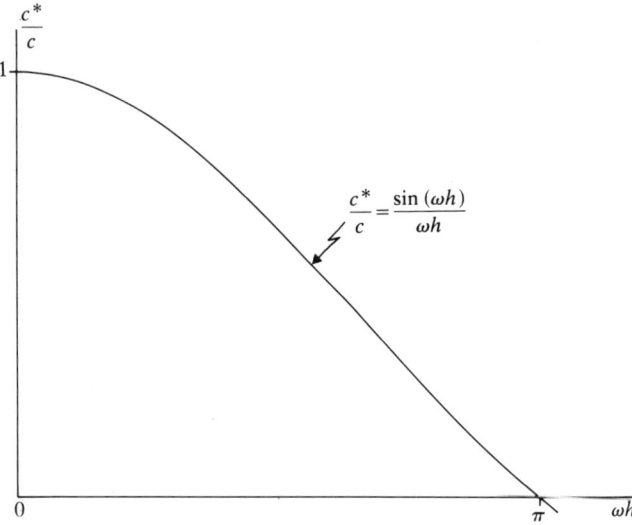

FIG. 2.1a. *Phase velocity of sinusoidal solutions of the explicit 3-point semi-discretization* (2.13).

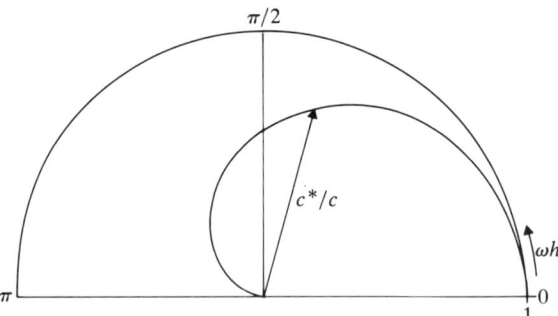

FIG. 2.1b. *Another way to display $c^*(\omega)$ is to use the polar form illustrated in this figure (same case as* FIG. 2.1a). *This form of display was used in Roberts and Weiss* (1966).

has the spectral function

$$\hat{A}(\omega) = -ic\omega\left(\frac{\tan(\omega h/2)}{\omega h/2}\right)$$

and the numerical phase velocity

(2.14′) $$c^*(\omega) = c\left(\frac{\tan(\omega h/2)}{\omega h/2}\right).$$

By contrast with the preceding example, we have here $c^*(\omega) > c$ for all ω. See Fig. 2.1c and Table 2.1.

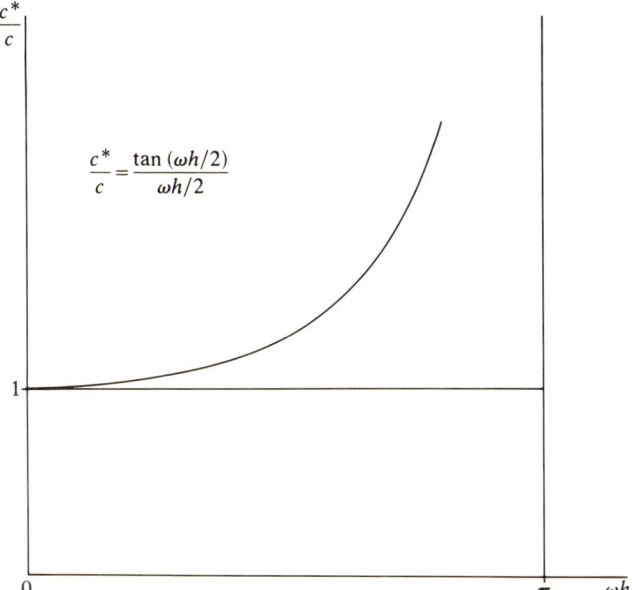

FIG. 2.1c. *Phase velocity of sinusoidal solutions of the implicit 2-point semi-discretization* (2.14).

2.3. Fourier transforms and global errors. A broader interpretation of the preceding analysis is obtained by replacing $V_\omega(t)$ with $\hat{U}(\omega, t)$ and $v_\omega(t)$ with $\bar{u}(\omega, t)$, that is, by replacing the somewhat heuristic concept of "sinusoidal trial solution" with the more powerful device of Fourier transforms. Other than this substitution, all the mathematics remain unchanged. A function $U(x, t)$ that is a solution of the advection equation (2.1) on $(-\infty, \infty)$ has the Fourier transform (from (2.8))

$$\hat{U}(\omega, t) = \hat{U}(\omega, 0) \, e^{-i\omega ct} \qquad (2.15a)$$

and may thus be expressed as

$$U(x, t) = \int_{-\infty}^{\infty} \hat{U}(\omega, 0) \, e^{i\omega(x-ct)} \frac{d\omega}{2\pi}. \qquad (2.15b)$$

Likewise, if $\{u_n\}$ is a semi-discrete approximation of U, it has the *discrete* Fourier transform

$$\bar{u}(\omega, t) = \bar{u}(\omega, 0) \, e^{\operatorname{Re}\hat{A}(\omega)t} \, e^{-i\omega c^*(\omega)t} \qquad (2.16a)$$

and may thus be expressed as

$$u_n(t) = \int_{-\pi/h}^{\pi/h} \bar{u}(\omega, 0) \, e^{\operatorname{Re}\hat{A}(\omega)t} \, e^{i\omega(x_n - c^*t)} \frac{d\omega}{2\pi}. \qquad (2.16b)$$

The exact solution (2.15) has the property of having a constant $\mathscr{L}_2$-norm. In certain engineering-related disciplines, the integral of the square of physical quantities is related to energy. By extension here it may be said that solutions of the advection equation on the entire real axis $(-\infty, \infty)$ are energy conservative:

$$\|U(x,t)\|_2^2 \equiv \int_{-\infty}^{\infty} |U(x,t)|^2 \, dx = \text{constant}. \tag{2.17}$$

Indeed, this is easily proven by invoking Parseval's equality (1.60). For a semi-discretization which is conservative in the sense that $\operatorname{Re} \hat{A}(\omega) = 0$ for all ω, we have preservation of the conservative property in the $\mathscr{L}_2$-norm, i.e., the numerical solution satisfies:

$$\begin{aligned}\|u_n\|_2^2 &\equiv h \sum_{n=-\infty}^{\infty} |u_n|^2 \\ &= \int_{-\pi/h}^{\pi/h} |\bar{u}(\omega, 0)|^2 \cdot |e^{i\omega(x-c^*t)}|^2 \frac{d\omega}{2\pi} = \text{constant}.\end{aligned} \tag{2.18}$$

The term "energy conservative" is thus used to describe semi-discretizations for which $\operatorname{Re} \hat{A}(\omega) \equiv 0$.

When $\operatorname{Re} \hat{A}(\omega) \leq 0$, and the $<$ sign holds for some ω, then, in general, $\|u_n\|_2^2$ decreases with time. *Dissipation* (of energy) is introduced *spuriously* by the approximation.

Spurious dispersion: Exact solutions of the advection equation are nondispersive. That is, the shape of the initial function is preserved:

$$U(x,t) = U(0, x-ct).$$

The nonconstancy of c^* in discrete approximations results in a nonconstant velocity of propagation of the harmonic components into which a nonsinusoidal initial function is resolved by Fourier analysis. This creates *spurious dispersion* in numerical solutions and nonsinusoidal functions lose their shape in their time evolution.

2.4. Relation to classical truncation error analysis. The classical definition of the truncation error of a semi-discretization is as follows:

Let $U(x,t)$ be a solution of the exact equation (2.1). Then the truncation error of the semi-discretization (2.2) is, by definition,

$$T_h = \frac{dU_n}{dt} - \mathbf{A} \cdot U_n, \tag{2.19}$$

where

$$\{U_n(t)\} = \{U(x_n, t)\}$$

are values of $U(x, t)$ taken on the lines $\{x = x_n\}$. The standard way to express (2.19) is by using Taylor series. For example, to obtain an expression for the truncation error of the simple explicit semi-discretization (1.19) we expand U in a Taylor series as

$$(2.20) \qquad U_{n\pm 1} = U_n \pm h\left(\frac{\partial U}{\partial x}\right)_n + \frac{h^2}{2}\left(\frac{\partial^2 U}{\partial x^2}\right)_n \pm \cdots.$$

Inserting (2.20) into (2.19) we find:

$$(2.21) \qquad T_h = \frac{dU_n}{dt} + c\left[\left(\frac{\partial U}{\partial x}\right)_n + \frac{h^2}{3}\left(\frac{\partial^3 U}{\partial x^3}\right)_n + \cdots\right].$$

Since $U_t + cU_x = 0$, this becomes

$$(2.22) \qquad T_h = c\frac{h^2}{3}\left(\frac{\partial^3 U}{\partial x^3}\right)_n + \text{higher order terms} = O(h^2).$$

All the semi-discretizations of $U_t + cU_x = 0$ which were described in § 1.4 have coefficients of $\mathbf{A} \cdot$ containing h^{-1}, and an expression of the truncation error which generalizes (2.22) for those expressions is

$$(2.23) \qquad T_h = Kh^p\left(\frac{\partial^{p+1} U}{\partial x^{p+1}}\right) + \text{higher order terms} = O(h^p)$$

where K is a constant independent of U and h, and p, an integer (which is even for all conservative semi-discretizations), is called the *order of accuracy* of the semi-discretization.

There is a direct relationship between order of accuracy and the "flatness" of $c^*(\omega)$ near $\omega = 0$. Indeed, by taking the Fourier transform of (2.19) and using (2.23), we obtain, for conservative semi-discretizations,

$$(2.24) \qquad \begin{aligned} \mathcal{F}(T_h) &= -ic\omega \hat{U}(\omega, t) - \hat{A}(\omega)\hat{U}(\omega, t) \\ &= -i\omega(c - c^*(\omega))\hat{U}(\omega, t) \\ &= Kh^p(i\omega)^{p+1}\hat{U}(\omega, t) + \text{higher order terms}. \end{aligned}$$

Hence

$$(2.25) \qquad \begin{aligned} c^*(\omega) - c &= Kh^p(i\omega)^p + \text{higher order terms} \\ &= O(\omega^p). \end{aligned}$$

The phase velocity error $c^*(\omega) - c$ behaves like ω^p near $\omega = 0$, or, in other words $c^*(\omega)$ has $p - 1$ zero derivatives in $\omega = 0$ (p is the order of accuracy of the semi-discretization).

The condition

$$(2.26) \qquad T_h = O(h^p), \quad p \geq 1$$

(called the *consistency condition* of the semi-discretization) becomes in the frequency domain

(2.27) $$c^*(\omega) - c = O(\omega^p), \quad p \geq 1,$$

where p is the same in both cases. The order of accuracy p is also the rate (expressed in terms of powers of h) at which the numerical solution converges to the exact solution. A formal derivation of this property is given in § 8.4, as a byproduct of a Fourier analysis of the global error.

2.5. Finite element Galerkin semi-discretizations. We shall now analyze the accuracy of the semi-discretization of $U_t + cU_x = 0$ obtained with the linear finite element Galerkin method (1.30), repeated here for convenience,

(2.28) $$\mathbf{A}_1 \cdot \frac{du_n}{dt} \equiv \frac{1}{6}\left(\frac{du_{n-1}}{dt} + 4\frac{du_n}{dt} + \frac{du_{n+1}}{dt}\right) = -c\left(\frac{u_{n+1} - u_{n-1}}{2h}\right) \equiv \mathbf{A}_2 \cdot u_n.$$

Following (2.6'), we derive

(2.29)
$$\hat{A}_1(\omega) = \frac{\mathbf{A}_1 \cdot e^{i\omega x_n}}{e^{i\omega x_n}} = \frac{2 + \cos(\omega h)}{3},$$

$$\hat{A}_2(\omega) = \frac{\mathbf{A}_2 \cdot e^{i\omega x_n}}{e^{i\omega x_n}} = -ic\left(\frac{\sin(\omega h)}{h}\right),$$

$$\hat{A}(\omega) = \frac{\hat{A}_2(\omega)}{\hat{A}_1(\omega)} = -ic\omega\left(\frac{3}{2 + \cos(\omega h)}\right)\left(\frac{\sin(\omega h)}{\omega h}\right).$$

The phase velocity of the approximation is thus found from (2.11) to be

(2.30) $$c^*(\omega) = c\left(\frac{3}{2 + \cos(\omega h)}\right)\left(\frac{\sin(\omega h)}{\omega h}\right).$$

This function is illustrated in Fig. 2.2 and summarized in Table 2.1.

Expanding (2.30) in a Taylor series near the origin gives

(2.31) $$c^*(\omega) = c(1 + O[(\omega h)^4]).$$

The truncation error in the approximation of the hyperbolic equation (2.1) with a linear finite element Galerkin semi-discretization is thus of order $O(h^4)$. Of all semi-discretizations of the form

(2.32) $$\sum_{k=-1}^{1} a_{1,k} \frac{du_{n+k}}{dt} = \sum_{k=-1}^{1} a_{2,k} u_{n+k},$$

the one given by (2.28) has the highest order of accuracy in the classical sense. As illustrated in Fig. 2.2, the approximation is superior to the simple central differences approximation in the entire frequency range, $\omega \in (0, \pi/h)$.

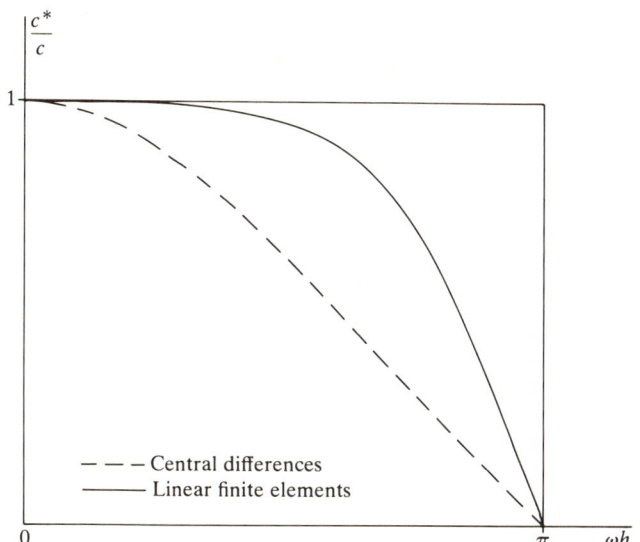

FIG. 2.2. *Phase velocity of sinusoidal solutions of the linear finite element/Galerkin implicit semi-discretization* (2.28). *The 3-point central differences semi-discretization* (Fig. 2.1a) *is shown for comparison.*

2.6. A generalization. We may express a family of 3-point semi-discretizations of the advection equation as

$$(2.33) \quad \left(\frac{\beta}{2}\frac{du_{n-1}}{dt} + (1-\beta)\frac{du_n}{dt} + \frac{\beta}{2}\frac{du_{n+1}}{dt}\right) = -c\left(\frac{u_{n+1}-u_{n-1}}{2h}\right),$$

where β is a free parameter. With $\beta = 0$ this is the central finite difference approximation. With $\beta = \frac{1}{3}$ it is the linear finite element/Galerkin approximation and for $\beta = \frac{1}{2}$ it is equivalent to the "box" method. The corresponding phase velocity is easily found to be

$$(2.34) \quad c^*(\omega) = c\left(\frac{1}{1-\beta+\beta\cos(\omega h)}\right)\left(\frac{\sin(\omega h)}{\omega h}\right).$$

The ratio c^*/c for various values of β is shown in Fig. 2.3 and summarized in Table 2.1.

Note that the general semi-discretization (2.33) may be obtained by application of a *method of weighted residuals*. With linear finite-elements and an equation residual defined as before by (1.25), the Galerkin conditions (1.26) are replaced by the *weighted residual* condition:

$$(2.35) \quad \langle w_n(x), \mathcal{R} \rangle \equiv \int w_n(x)\mathcal{R}\,dx = 0, \quad n = \cdots, 0, 1, 2, \cdots.$$

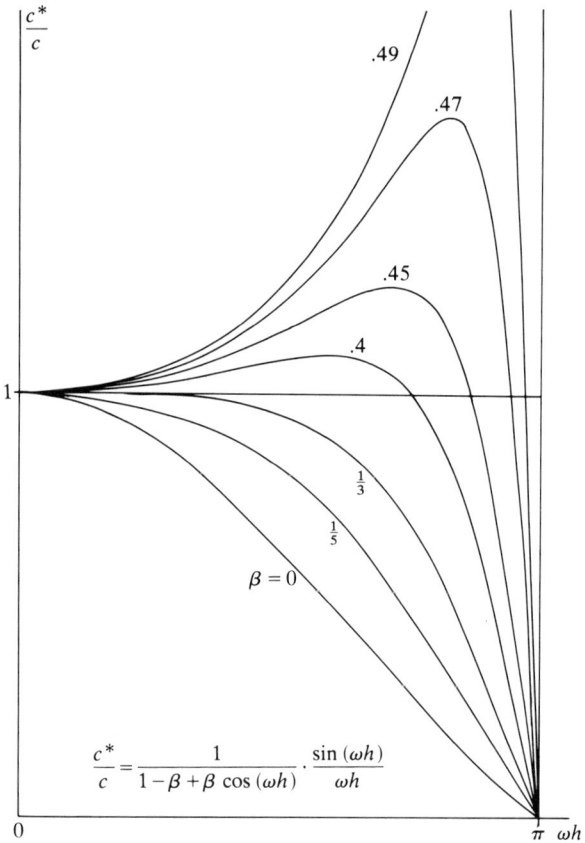

FIG. 2.3. *Phase velocity for the general 3-point family of semi-discretizations of $U_t + cU_x = 0$ for various values of β.*

The weight functions are defined as (see FIG. 2.4)

$$(2.36) \qquad w_n(x) = \begin{cases} \dfrac{1}{2\gamma h} & \text{for } |x - x_n| \leq \gamma h, \\ 0 & \text{elsewhere.} \end{cases}$$

Then we find that

$$(2.37) \qquad \frac{\gamma}{4} \frac{du_{n-1}}{dt} + \left(1 - \frac{\gamma}{2}\right) \frac{du_n}{dt} + \frac{\gamma}{4} \frac{du_{n+1}}{dt} = -c\left(\frac{u_{n+1} - u_{n-1}}{2h}\right),$$

which is precisely (2.33) with $\beta = \gamma/2$.

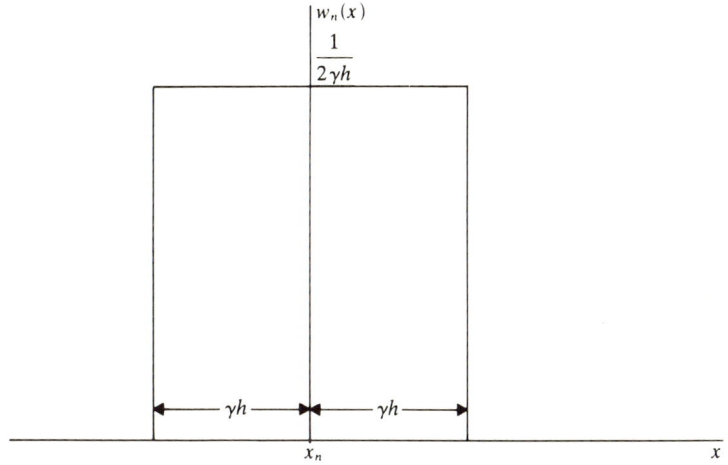

FIG. 2.4. *Weight function for the method of weighted residuals.*

2.7. Semi-discretization of a conservation law. The general expression of a hyperbolic system written in *conservation law* form[3] is

$$\frac{\partial U}{\partial t} + \frac{\partial}{\partial x}(F(U)) = 0. \tag{2.38}$$

By integration between two successive grid points we find the *exact* relation

$$\frac{d}{dt}\int_n^{n+1} U(x,t)\,dx + F(U_{n+1}) - F(U_n) = 0. \tag{2.39}$$

If U is approximated with linear finite elements over the same grid points, then (2.39) becomes (after division by h) the 2-point semi-discretization

$$\frac{1}{2}\left(\frac{du_n}{dt} + \frac{du_{n+1}}{dt}\right) + \frac{F(u_{n+1}) - F(u_n)}{h} = 0 \tag{2.40}$$

or, for our simple model equation (2.1), the expression which we have used before,

$$\frac{1}{2}\left(\frac{du_n}{dt} + \frac{du_{n+1}}{dt}\right) + c\left(\frac{u_{n+1} - u_n}{h}\right) = 0. \tag{2.41}$$

There is of course a great degree of similarity between this approach and the method of weighted residuals just described. For example, (2.41) is equivalent to (2.33) when $\beta = \frac{1}{2}$.

Table 2.1 summarizes the semi-discretizations discussed in §§ 2.1–2.7.

---
[3] See Lax and Wendroff (1960).

TABLE 2.1
Semi-discretizations of $du/dt + cdu/dx = 0$

Semi-discretization	$\hat{A}(\omega)$	$c^*(\omega)$
Simple 3-point finite differences		
$\dfrac{du_n}{dt} = -c\left(\dfrac{u_{n+1}-u_{n-1}}{2h}\right)$	$-ic\dfrac{\sin(\omega h)}{h}$	$c\dfrac{\sin(\omega h)}{\omega h}$
2-point implicit ("box") ($\beta = \frac{1}{2}$)		
$\dfrac{1}{2}\left(\dfrac{du_n}{dt}+\dfrac{du_{n+1}}{dt}\right)$ $= -c\left(\dfrac{u_{n+1}-u_n}{h}\right)$	$-ic\dfrac{\tan(\omega h/2)}{h/2}$	$c\dfrac{\tan(\omega h/2)}{\omega h/2}$
Linear finite element Galerkin ($\beta = \frac{1}{3}$)		
$\dfrac{1}{6}\dfrac{du_{n-1}}{dt}+\dfrac{4}{6}\dfrac{du_n}{dt}+\dfrac{1}{6}\dfrac{du_{n+1}}{dt}$ $= -c\left(\dfrac{u_{n+1}-u_{n-1}}{2h}\right)$	$\dfrac{-3ic\sin(\omega h)}{h(2+\cos(\omega h))}$	$\dfrac{3c\sin(\omega h)}{\omega h(2+\cos(\omega h))}$
General 3-point family		
$\dfrac{\beta}{2}\dfrac{du_{n-1}}{dt}+(1-\beta)\dfrac{du_n}{dt}+\dfrac{\beta}{2}\dfrac{du_{n+1}}{dt}$ $= -c\left(\dfrac{u_{n+1}-u_{n-1}}{2h}\right)$	$\dfrac{-ic\sin(\omega h)}{h(1-\beta+\beta\cos(\omega h))}$	$\dfrac{c\sin(\omega h)}{\omega h(1-\beta+\beta\cos(\omega h))}$

2.8. Time frequency. The relationship between time frequency Ω and space frequency ω for any sinusoidal wave is simply

$$\text{time frequency} = \text{space frequency} * \text{velocity}$$

or

(2.42) $\qquad \Omega = \omega \times \text{velocity}.$

For exact sinusoidal solutions of $U_t + cU_x = 0$ we have

(2.43) $\qquad \Omega = \omega c.$

(Note that this is simply the coefficient of t in the exponent of (2.8).)
For numerical sinusoidal solutions this becomes

(2.44) $\qquad \Omega^* = \omega c^*(\omega).$

ANALYSIS OF THE ACCURACY OF SEMI-DISCRETIZATIONS

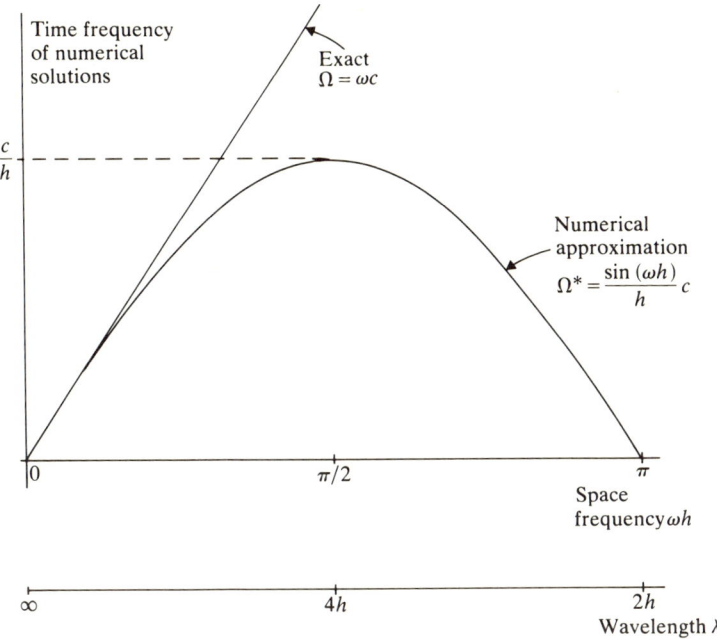

FIG. 2.5. *Relation between the time frequency Ω and space frequency ω for the exact solution of $U_t + cU_x = 0$ and the numerical approximation using the central differences semi-discretization (2.13).*

In the simple central differences case, we have

$$\Omega^* = c\left(\frac{\sin(\omega \cdot h)}{h}\right). \tag{2.45}$$

This function is illustrated in Fig. 2.5 along with the exact relation (2.43) for comparison. We see that in this case no numerical sinusoidal components can exist for

$$\Omega^* > \frac{c}{h} \quad \text{or} \quad \frac{\Omega^* h}{c} > 1.$$

Further aspects of this question are developed in Chapter 7.

2.9. The wave equation. Numerical approximations of the wave equation

$$\frac{\partial^2 U}{\partial t^2} = c^2 \frac{\partial^2 U}{\partial x^2} \tag{2.46}$$

can be obtained without transforming it to an equivalent system of first order equations. The semi-discretization

(2.47) $$\frac{d^2 u_n}{dt^2} = c^2 \left(\frac{u_{n-1} - 2u_n + u_{n+1}}{h^2} \right),$$

which contains a second derivative with respect to time, may be integrated as

(2.48) $$u_n^{j+1} = -u_n^{j-1} + 2u_n^j + \left(\frac{c \, \Delta t}{h} \right)^2 (u_{n-1}^j - 2u_n^j + u_{n+1}^j).$$

This formula was used in Courant–Friedrichs–Lewy (1928) and is the original discretization for which the stability condition

(2.49) $$\left(\frac{c \, \Delta t}{h} \right) \leq 1$$

was derived. ($c \, \Delta t / h$ is called the *Courant number*.)

To (2.48) are associated the two characteristic velocities

$$\nu = \pm c.$$

These are also the velocities at which (exact) sinusoidal solutions propagate. It is easy to verify that

(2.50) $$U_\omega(x, t) = V_\omega(0) \, e^{i\omega(x \pm ct)}$$

is indeed a solution of (2.46).

In a manner similar to that used before, we may seek the velocity at which sinusoidal solutions of the semi-discretization (2.47) propagate. Such solutions are of the general form

(2.51) $$u_\omega(x_n, t) = v_\omega(t) \, e^{i\omega x_n}.$$

Upon insertion into (2.47), we find for $v_\omega(t)$ that

(2.52) $$\frac{d^2 v_\omega}{dt^2} = c^2 \left[\frac{2 \cos(\omega h) - 2}{h^2} \right] v_\omega$$

$$= -c^2 \left[\frac{\sin(\omega h/2)}{\omega h/2} \right]^2 \omega^2 v_\omega$$

or, by analytic integration,

(2.53) $$v_\omega(t) = v_\omega(0) \, e^{\pm i c \omega [\sin(\omega h/2)/(\omega h/2)] t}.$$

That is, (2.51) becomes

(2.54) $$u_\omega(x_n, t) = v_\omega(0) \, e^{i\omega(x_n \pm c^*(\omega) t)},$$

where

(2.55) $$c^*(\omega) = c\left(\frac{\sin(\omega h/2)}{\omega h/2}\right)$$

is the corresponding numerical phase velocity.

We may observe that this numerical velocity is as if the wave equation had been transformed into a system of first order equations and then semi-discretized over a mesh of size $h/2$. Note that this result could have been inferred a priori from the operator identity (see also § 3.1 below)

(2.56) $$\boldsymbol{D}_2 \cdot \equiv \boldsymbol{E}^{-1} \cdot - 2 + \boldsymbol{E} \cdot = \delta^2 \cdot,$$

where

(2.57) $$\delta \cdot \equiv \boldsymbol{E}^{1/2} \cdot - \boldsymbol{E}^{-1/2} \cdot.$$

Indeed, the corresponding relations between spectral functions are

(2.58) $$-\omega^2 h^2 \simeq \hat{D}_2 = (\hat{\delta})^2 = -\left(2\sin\left(\frac{\omega h}{2}\right)\right)^2.$$

2.10. Implicit semi-discretizations of the wave equation. A general form of implicit semi-discretization of the wave equation analogous to that in (2.33) is

(2.59) $$\frac{\beta}{2}\ddot{u}_{n-1} + (1-\beta)\ddot{u}_n + \frac{\beta}{2}\ddot{u}_{n+1} = c^2\left(\frac{u_{n-1} - 2u_n + u_{n+1}}{h^2}\right).$$

The value $\beta = \frac{1}{3}$ corresponds to the semi-discretization obtained with linear finite elements. The value $\beta = \frac{1}{6}$ is that which endows (2.59) with the highest order of accuracy. The corresponding expression

(2.60) $$\tfrac{1}{12}(\ddot{u}_{n-1} + 10\ddot{u}_n + \ddot{u}_{n+1}) = c^2\left(\frac{u_{n-1} - 2u_n + u_{n+1}}{h^2}\right)$$

is an application of the Störmer–Numerov formula[4] and has the truncation error

$$T_h = \tfrac{1}{12}(\ddot{U}_{n-1} + 10\ddot{U}_n + \ddot{U}_{n+1}) - \frac{c^2}{h^2}(U_{n-1} - 2U_n + U_{n+1})$$

(2.61) $$= \tfrac{1}{240} c^2 h^4 \left(\frac{\partial^6 U}{\partial x^6}\right) + \text{higher order terms}$$

$$= O(h^4)$$

---

[4] See, e.g., Birkhoff and Gulati (1974).

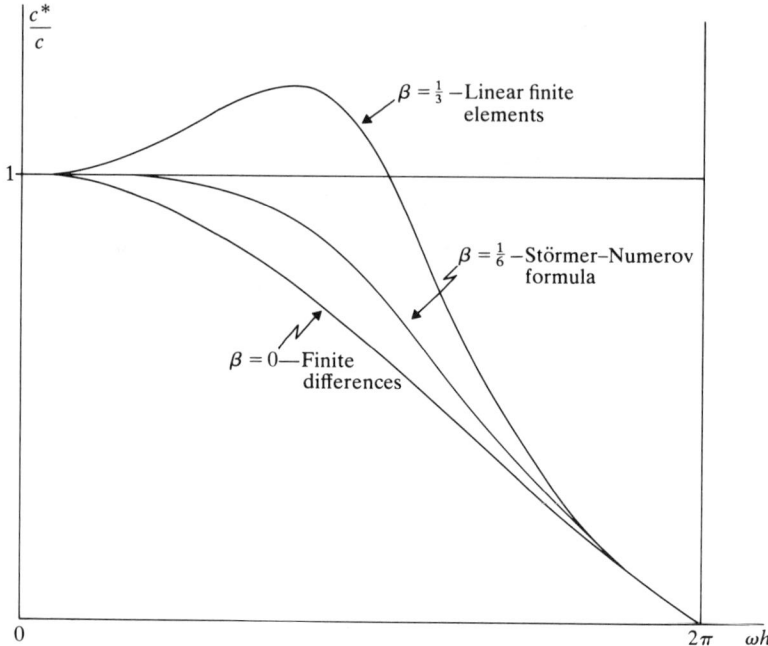

FIG. 2.6. *Phase velocity of implicit semi-discretizations of the wave equation $U_{tt} = c^2 U_{xx}$.*

(all other semi-discretizations of the form (2.59) have a truncation error which is $O(h^2)$).

Proceeding as before, the expression for the phase velocity of the semi-discretization (2.5) is found to be (see Fig. 2.6)

$$(2.62) \qquad c^*(\omega) = c\left(\frac{\sin(\omega h/2)}{\omega h/2}\right)(1 - \beta + \beta \cos(\omega h))^{-1/2}.$$

Note the fact that with linear finite elements ($\beta = \frac{1}{3}$) we have $c^*(\omega) > c$ for $0 < |\omega| \leq \pi/c$. That is, the numerical approximation "stiffens" the propagating medium.[5] By contrast, $c^* < c$ with simple finite differences ($\beta = 0$) making the propagation properties "looser" in that case. (What happens to $c^*(\omega)$ when $|\omega| > \pi/h$ is of no concern, since numerical solutions are entirely described by $\bar{u}(\omega, t)$ with $|\omega| \in [0, \pi/h]$; see, in particular § 1.12.)

---

[5] See Birkhoff and Dougalis (1975).

Chapter 3

Higher Order Semi-Discretizations

3.1. Synthesis in the frequency domain. The fact that order of accuracy and the flatness of $c^*(\omega)$ near $\omega = 0$ are both maximized by using free coefficients in higher order finite difference approximations to

$$\frac{\partial U}{\partial t} + c\frac{\partial U}{\partial x} = 0 \tag{3.1}$$

may be used as a tool for synthesis. We seek approximations to $\boldsymbol{L}\cdot \equiv \partial/\partial x$ of the form

$$\boldsymbol{L}\cdot \simeq \boldsymbol{A}\cdot \equiv \sum_{k=-K}^{K} a_k \boldsymbol{E}^k \cdot . \tag{3.2}$$

These will be even order of accuracy approximations, and the conservative property $\operatorname{Re} \hat{A}(\omega) = 0$ for all ω implies anti-symmetry, $a_k = -a_{-k}$.

We define the basic difference operators

$$\boldsymbol{D}_1 \cdot = \frac{\boldsymbol{E}\cdot - \boldsymbol{E}^{-1}\cdot}{2}, \tag{3.3}$$

$$\boldsymbol{D}_2 \cdot = \boldsymbol{E}\cdot - 2 + \boldsymbol{E}^{-1}\cdot \tag{3.4}$$

and will verify that approximations of the form (3.2) can be obtained in the form

$$\boldsymbol{A}\cdot = \boldsymbol{D}_1 \cdot \sum_{r=0}^{K-1} \alpha_r \boldsymbol{D}_2^r \cdot . \tag{3.5}$$

That such approximations can indeed be obtained shall be demonstrated a posteriori by the fact that the Taylor series of the spectral function $\hat{A}(\omega)$ and $\hat{L}(\omega) = i\omega$ can be made to agree, term by term, up to the term in ω^{2K}.

We find the spectral function of the operators $\boldsymbol{D}_1\cdot$ and $\boldsymbol{D}_2\cdot$ respectively:

$$\hat{D}_1(\omega) = i \sin(\omega h), \tag{3.6}$$

$$\hat{D}_2(\omega) = 2\cos(\omega h) - 2 = -\left(2\sin\left(\frac{\omega h}{2}\right)\right)^2. \tag{3.7}$$

Thus, $\hat{A}(\omega)$ is found:

$$\hat{A}(\omega) = i \sin(\omega h) \sum_{r=0}^{K-1} \alpha_r (-1)^r \left(2 \sin\left(\frac{\omega h}{2}\right)\right)^{2r}. \tag{3.8}$$

Accordingly, the operator approximation

$$\boldsymbol{L} \cdot \simeq \boldsymbol{D}_1 \cdot \sum_{r=0}^{K-1} \alpha_r \boldsymbol{D}_2^r \cdot \tag{3.9}$$

becomes in the frequency domain

$$i\omega \simeq i \sin(\omega h) \sum_{r=0}^{K-1} \alpha_r (-1)^r \left(2 \sin\left(\frac{\omega h}{2}\right)\right)^{2r} \tag{3.10}$$

or

$$\frac{\omega h}{\sin(\omega h)} \simeq h \sum_{r=0}^{K-1} \alpha_r (-1)^r \left(2 \sin\left(\frac{\omega h}{2}\right)\right)^{2r}. \tag{3.11}$$

From the tabulated expansion[1]

$$\frac{\theta}{\sin \theta} = 1 + \frac{2}{3} \sin^2\left(\frac{\theta}{2}\right) + \frac{2 \cdot 4}{3 \cdot 5} \sin^4\left(\frac{\theta}{2}\right) + \cdots \tag{3.12}$$

we obtain simply, by letting $\omega h = \theta$,

$$\alpha_0 = \frac{1}{h}, \tag{3.13}$$

$$\alpha_1 = -\frac{1}{h} \cdot \frac{2}{3} \cdot \frac{1}{2^2} = -\frac{1}{6h}, \tag{3.14}$$

$$\alpha_2 = \frac{1}{h} \cdot \frac{2 \cdot 4}{3 \cdot 5} \cdot \frac{1}{2^4} = \frac{1}{30h}, \cdots \tag{3.15}$$

That is,

$$\frac{\partial}{\partial x} \simeq \boldsymbol{A} \cdot = \frac{\boldsymbol{D}_1 \cdot}{h} (1 - \tfrac{1}{6}\boldsymbol{D}_2 \cdot + \tfrac{1}{30}\boldsymbol{D}_2^2 \cdots). \tag{3.16}$$

Values of a_k so obtained are tabulated in Table 3.1. These are of course the unique set of coefficients giving higher order finite difference approximations of $\boldsymbol{L} \cdot = \partial/\partial x$, independently of how they were derived.

The interesting property of the approximation with the coefficients $\{\alpha_r\}$ given by (3.13)–(3.15) is that increasing the approximation by two orders of accuracy is obtained by adding one more term to the series, *without change to those already there*. By contrast, when the approximation is written in the

---

[1] See Jolley (1925, p. 116).

TABLE 3.1
Coefficients of central difference approximations of the first derivative

Number of points	Coefficients a_k								
	a_{-4}	a_{-3}	a_{-2}	a_{-1}	a_0	a_1	a_2	a_3	a_4
3				$\dfrac{-1}{2h}$	0	$\dfrac{1}{2h}$			
5			$\dfrac{1}{12h}$	$\dfrac{-8}{12h}$	0	$\dfrac{8}{12h}$	$\dfrac{-1}{12h}$		
7		$\dfrac{-1}{60h}$	$\dfrac{9}{60h}$	$\dfrac{-45}{60h}$	0	$\dfrac{45}{60h}$	$\dfrac{-9}{60h}$	$\dfrac{1}{60h}$	
9	$\dfrac{3}{840h}$	$\dfrac{-32}{840h}$	$\dfrac{168}{840h}$	$\dfrac{-672}{840h}$	0	$\dfrac{672}{840h}$	$\dfrac{-168}{840h}$	$\dfrac{32}{840h}$	$\dfrac{-3}{840h}$

equivalent form (3.2), obtained for instance by the method of undetermined coefficients, then whenever K is increased, *all the coefficients a_k must be recalculated* (see Table 3.1).

3.2. Velocity error. Consider now the family of explicit semi-discretizations of $U_t + cU_x = 0$, where $\partial/\partial x$ has been approximated by the higher order expressions (3.16). Using the expressions (2.11) for the corresponding phase velocity, we readily find

$$(3.17) \quad c^*(\omega) = c \frac{\sin(\omega h)}{\omega h}\left[1 + \frac{2}{3}\sin^2\left(\frac{\omega h}{2}\right) + \frac{8}{15}\sin^4\left(\frac{\omega h}{2}\right) + \cdots\right]$$

where the expansion is to be terminated after the appropriate number of terms. These functions are illustrated in Fig. 3.2.

3.3. Limiting case ($K \to \infty$). The trigonometric series (3.8) is convergent for $|\omega h| < \pi$. In fact, it converges on that interval to $i\omega$ as $K \to \infty$. Because of the nature of its terms, which are all periodic of period $2\pi/h$, the function defined by (3.8) over the whole real axis is simply that illustrated in Fig. 3.3.

3.4. Relation to the cardinal function. Suppose that we were to derive an approximation to $(\partial U/\partial x)_n$ by the following process:
(i) We construct the function

$$(3.18) \quad u^*(x) = \sum_{n=-\infty}^{\infty} \Psi_n(x) u_n,$$

where the $\Psi_n(x)$ are the cardinal functions defined by (1.64).

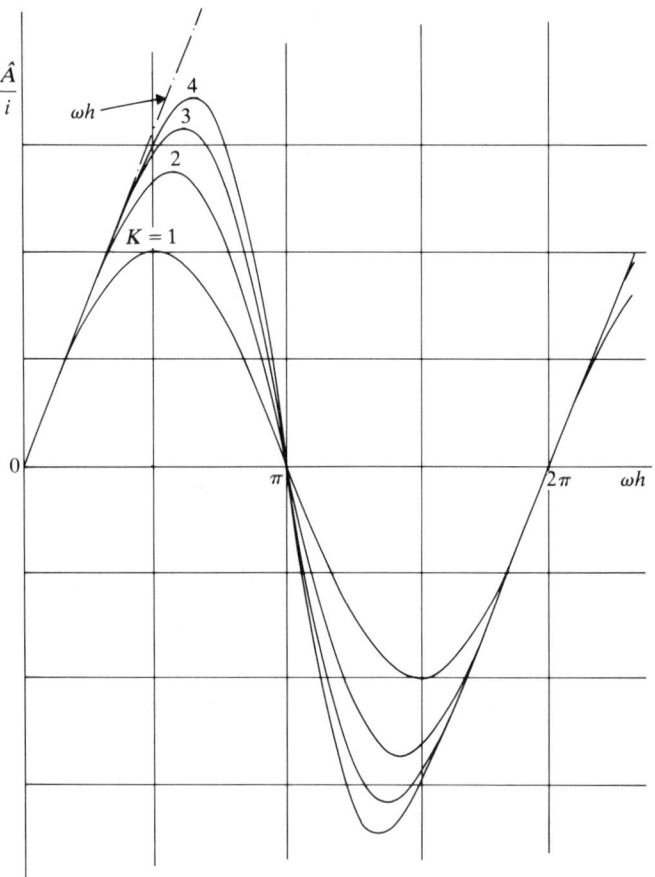

FIG. 3.1. *Approximation* (3.10) *for increasing values of K.*

(ii) We approximate $(\partial U/\partial x)_n$ by

(3.19) $$\left(\frac{\partial U}{\partial x}\right)_n \simeq \left(\frac{\partial u^*}{\partial x}\right)_n = \sum_m u_m \left(\frac{d\Psi_m}{dx}\right)_n.$$

If we let

(3.20) $$b_m = \left(\frac{d\Psi_0}{dx}\right)_m$$

then (3.19) may be expressed simply as

(3.21) $$\left(\frac{\partial U}{\partial x}\right)_n \simeq \sum_{m=-\infty}^{\infty} b_m u_{n+m}.$$

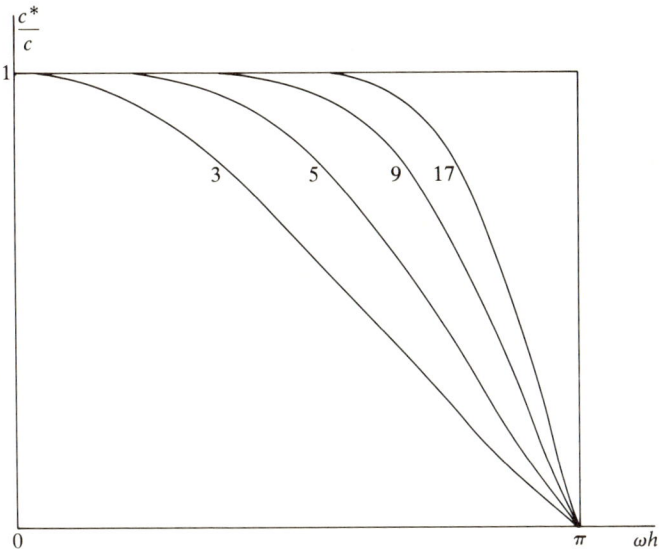

FIG. 3.2. *Phase velocity for higher order semi-discretizations with 3, 5, 9 and 17 points (1, 2, 4 and 8 terms in the series (3.17)).*

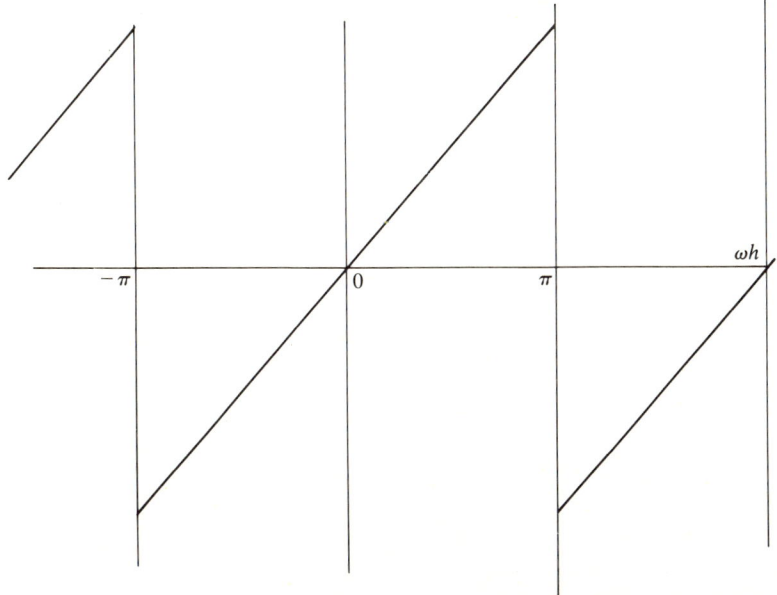

FIG. 3.3. *The function $\hat{A}(\omega)/i$ when $K \to \infty$.*

As it turns out, the approximation so obtained is identical to (3.16) when $K \to \infty$. The proof is rather straightforward and shall be omitted here.

3.5. B-spline Galerkin semi-discretizations. We describe now an interesting method for the derivation of higher order implicit semi-discretizations which is based on the use of B-splines.

Consider the B-spline Galerkin semi-discretizations of $U_t + cU_x = 0$ that were derived in § 1.6, viz.,

$$(3.22) \qquad \frac{1}{h} \sum_m \langle \varphi_n^q, \varphi_m^q \rangle \frac{dv_m}{dt} = -\frac{c}{h} \sum_m \left\langle \varphi_n^q, \frac{d\varphi_m^q}{dx} \right\rangle v_m,$$

where $\{\varphi_n^q(x)\}$ is the basis of B-splines of degree q defined by (1.39), and the $\{v_n(t)\}$ are the coefficients in the expressions of $u(x, t)$ in that basis. This equation may also be written as

$$(3.22') \qquad \frac{1}{h} \sum_k \langle \varphi_0^q, \varphi_k^q \rangle \frac{dv_{n+k}}{dt} = -\frac{c}{h} \sum_k \left\langle \varphi_0^q, \frac{d\varphi_k^q}{dx} \right\rangle v_{n+k}$$

or, in operator notation, as

$$(3.23) \qquad \boldsymbol{A}_1 \cdot \frac{dv_n}{dt} = \boldsymbol{A}_2 \cdot v_n,$$

where $\boldsymbol{A}_1 \cdot$ and $\boldsymbol{A}_2 \cdot$ are the operators

$$(3.24) \qquad \boldsymbol{A}_1 \cdot \equiv \frac{1}{h} \sum_k \langle \varphi_0^q, \varphi_k^q \rangle \boldsymbol{E}^k \cdot,$$

$$(3.25) \qquad \boldsymbol{A}_2 \cdot \equiv -\frac{c}{h} \sum_k \left\langle \varphi_0^q, \frac{d\varphi_k^q}{dx} \right\rangle \boldsymbol{E}^k \cdot,$$

The operators $\boldsymbol{A}_1 \cdot$ and $\boldsymbol{A}_2 \cdot$ are Toeplitz operators. Because of the finite support property of the $\{\varphi_n^q\}$, they are banded, i.e., they have nonzero coefficients only for $|k| \le q$.

3.6. An equivalence of bases. The algorithms (3.23) may be difficult to use in practice, since the calculations are in terms of the coefficients $\{v_n\}$ not the nodal values $\{u_n\}$. But this difficulty can be removed as follows: The relationship between the $\{u_n\}$ and the $\{v_n\}$ is expressed by (1.41'), viz.,

$$(3.26) \qquad u_n = \boldsymbol{T} \cdot v_n$$

where $\boldsymbol{T} \cdot$ is the Toeplitz operator

$$(3.27) \qquad \boldsymbol{T} \cdot \equiv \sum_k \varphi_0^q(kh) \boldsymbol{E}^k \cdot.$$

Multiplication of (3.23) by $T\cdot$ yields

(3.28) $$T\cdot A_1 \cdot \frac{dv_n}{dt} = T\cdot A_2 \cdot v_n.$$

But $T\cdot$, $A_1\cdot$ and $A_2\cdot$ commute (see § 1.5), whence

(3.29) $$A_1 \cdot \frac{d}{dt}(T\cdot v_n) = A_2 \cdot (T\cdot v_n)$$

or

(3.30) $$A_1 \cdot \frac{du_n}{dt} = A_2 \cdot u_n,$$

where $A_1\cdot$ and $A_2\cdot$ are the same as in (3.23). *This establishes that the nodal values $\{u_n\}$ satisfy the same evolution equation as the coefficients $\{v_n\}$ do.*

A procedure for the synthesis of higher order *implicit* semi-discretizations thus consists in deriving them using the B-spline Galerkin method, and in then using the resulting operators $A_1\cdot$ and $A_2\cdot$ in (3.30).

(This procedure as applied to the conservation law equation $U_t + [F(U)]_x = 0$, was described by Chin, Hedstrom and Karlsson (1979).)

Given below are the resulting algorithms for $q = 1, 2$ and 3.

$q = 1$. *Linear splines.*

(3.31a) $$\frac{1}{6}[4\dot{u}_n + (\dot{u}_{n+1} + \dot{u}_{n-1})] = -\frac{c}{2h}(u_{n+1} - u_{n-1}).$$

$q = 2$. *Quadratic splines.*

(3.31b) $$\frac{1}{120}[66\dot{u}_n + 26(\dot{u}_{n+1} + \dot{u}_{n-1}) + (\dot{u}_{n+2} + \dot{u}_{n-2})]$$
$$= -\frac{c}{24h}[10(u_{n+1} - u_{n-1}) + (u_{n+2} - u_{n-2})].$$

$q = 3$. *Cubic splines.*

(3.31c) $$\frac{1}{5040}[2416\dot{u}_n + 1191(\dot{u}_{n+1} + \dot{u}_{n-1}) + 120(\dot{u}_{n+2} + \dot{u}_{n-2}) + (\dot{u}_{n+3} + \dot{u}_{n-3})]$$
$$= -\frac{c}{720h}[245(u_{n+1} - u_{n-1}) + 56(u_{n+2} - u_{n-2}) + (u_{n+3} - u_{n-3})].$$

3.7. Equivalence with collocation. An interesting equivalence between B-spline Galerkin and B-spline collocation approximations was pointed out by Swartz and Wendroff (1974b). Because of the definition of B-splines as the successive convolutions of the characteristic function M_1 with itself, the following identities are true:

(3.32) $$\frac{1}{h}\langle \varphi_0^q, \varphi_k^q \rangle = \frac{1}{h}\int \varphi_0^q(x)\varphi_0^q(kh-x)\,dx = \varphi_0^{2q+1}(kh) = \varphi_k^{2q+1}(0)$$

and

(3.33) $$\frac{1}{h}\left\langle \varphi_0^q, \frac{d}{dx}\varphi_k^q \right\rangle = \frac{-1}{h}\int \varphi_0^q(x)\frac{d}{dx}\varphi_0^q(kh-x)\,dx$$
$$= -\frac{d}{dx}\varphi_0^{2q+1}(kh) = \frac{d}{dx}\varphi_k^{2q+1}(0).$$

Thus (3.22′) can also be expressed as

(3.34a) $$\sum_k \varphi_k^{2q+1}(0)\frac{dv_{n+k}}{dt} = -c\sum_k \frac{d}{dx}\varphi_k^{2q+1}(0)v_{n+k}$$

or

(3.34b) $$\sum_m \varphi_m^{2q+1}(x_n)\frac{dv_m}{dt} = -c\sum_m \frac{d}{dx}\varphi_m^{2q+1}(x_n)v_m.$$

If we had approximated the same problem with B-splines of degree $2q+1$ using collocation instead of Galerkin's method, then we would have written

(3.35) $$U(x,t) \simeq u(x,t) = \sum_n w_n(t)\varphi_n^{2q+1}(x)$$

and (collocation)

(3.36) $$\sum_m \varphi_m^{2q+1}(x_n)\frac{dw_m}{dt} = -c\sum_m \frac{d}{dx}\varphi_m^{2q+1}(x_n)w_m.$$

Thus, the evolution equations (3.34) for the $\{v_n\}$ (the degree q, B-spline Galerkin coefficients of u) are identical to the evolution equations (3.36) for the $\{w_n\}$ (the degree $(2q+1)$, B-spline collocation coefficients of u).

By the same argument as that used in § 3.6, the following evolution equations also apply:

(3.37) $$\sum_m \varphi_m^{2q+1}(x_n)\frac{du_m}{dt} = -c\sum_m \frac{d}{dx}\varphi_m^{2q+1}(x_n)u_m,$$

where the $\{u_n(t)\}$ are the nodal values of $u(x,t)$, and $u(x,t)$ is the spline interpolant of degree $2q+1$, with knots located at the nodes. The operators $A_1\cdot$ and $A_2\cdot$ defined in (3.24)–(3.25) may also be expressed as

(3.38) $$A_1\cdot = \sum_k \varphi_0^{2q+1}(kh)E^k\cdot = \sum_k \varphi_k^{2q+1}(0)E^k\cdot,$$

(3.39) $$A_2\cdot = c\sum_k \frac{d}{dx}\varphi_0^{2q+1}(kh)E^k\cdot = -c\sum_k \frac{d}{dx}\varphi_k^{2q+1}(0)E^k\cdot.$$

Summarizing previous results, we thus observe the interesting fact that *identical evolution equations* apply to:
  (i) the coefficients $\{v_n(t)\}$ of a qth degree B-spline Galerkin approximation of solutions of $U_t + cU_x = 0$;
  (ii) the nodal values $\{u_n(t)\}$ of the same spline;
  (iii) the coefficients $\{w_n(t)\}$ of a $(2q+1)$th degree B-spline collocation approximation;
  (iv) the nodal values $\{u_n(t)\}$ of that spline.

3.8. Fourier analysis of the algorithms obtained with B-splines. We now proceed with an analysis of the errors inherent in the algorithms (3.23). Their conservative property is evident from their symmetry: $\mathbf{A}_1 \cdot$ is symmetric and $\mathbf{A}_2 \cdot$ is antisymmetric. The errors are thus entirely characterized by the velocity error. From the definition of the numerical phase velocity

$$(3.40) \qquad c^*(\omega) = -\frac{1}{i\omega} \frac{\hat{A}_2(\omega)}{\hat{A}_1(\omega)},$$

we obtain, using the coefficients $\{a_{1,k}\}$, and $\{a_{2,k}\}$ of (3.31),

$q = 1$. *Linear splines.*

$$(3.41a) \qquad c^* = c \frac{3 \sin(\omega h)}{2 + \cos(\omega h)}.$$

$q = 2$. *Quadratic splines.*

$$(3.41b) \qquad c^* = c \frac{5[10 \sin(\omega h) + \sin(2\omega h)]}{33 + 26 \cos(\omega h) + \cos(2\omega h)}.$$

$q = 3$. *Cubic splines.*

$$(3.41c) \qquad c^* = c \frac{7[245 \sin(\omega h) + 56 \sin(2\omega h) + \sin(3\omega h)]}{1208 + 1191 \cos(\omega h) + 120 \cos(2\omega h) + \cos(3\omega h)}.$$

These are shown in Fig. 3.4.

3.9. Convergence rates. Convergence rates when $h \to 0$ may be analyzed by using the results of Fourier analysis as a starting point (Strang (1971), Thomée (1973), Swartz and Wendroff (1974a)).

We express $\hat{A}_1(\omega)$ and $\hat{A}_2(\omega)$ from (3.38)–(3.39) as

$$(3.42) \qquad \hat{A}_1(\omega) = \sum_k \varphi_0^{2q+1}(kh)\, e^{i\omega kh} = \frac{1}{h} \bar{\varphi}_0^{2q+1}(\omega),$$

$$(3.43) \qquad \hat{A}_2(\omega) = c \sum_k \frac{d\varphi_0^{2q+1}(kh)}{dx} e^{i\omega kh} = \frac{-c}{h} \overline{\frac{d\varphi_0^{2q+1}}{dx}}(\omega).$$

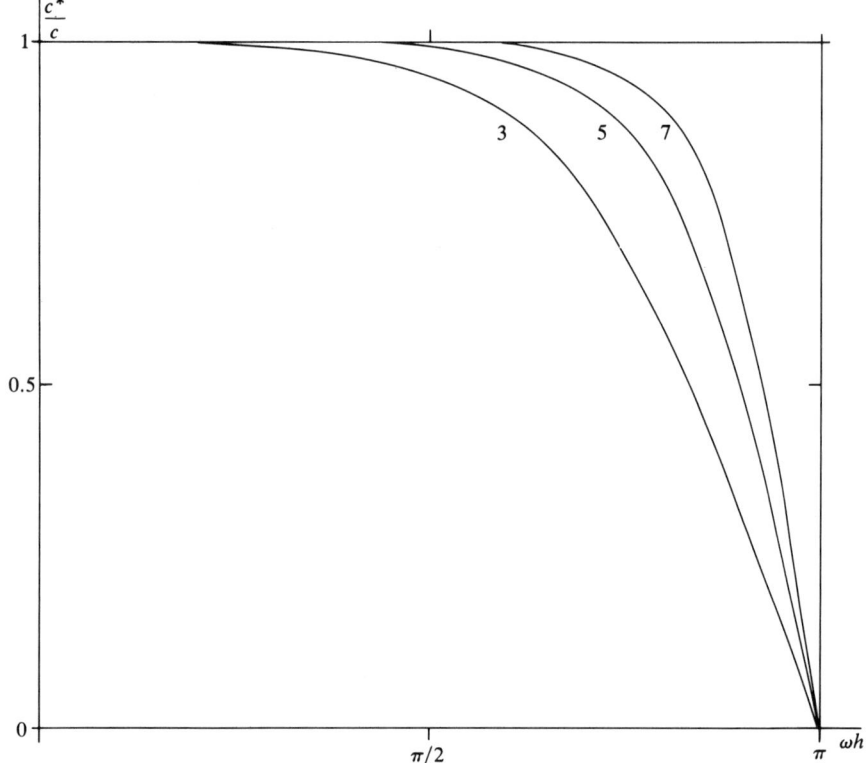

FIG. 3.4. *Phase velocity of higher order B-spline approximations with* 3, 5 *and* 7 *points (compare with Fig. 3.2).*

That is, they are, to within a simple factor, the discrete Fourier transforms of the B-spline $\varphi_0^{2p}(x)$ and of its derivative $d\varphi_0^{2q+1}/dx$. We note at this point the simplicity of the Fourier transform of B-splines. From the definition (1.36) of $M_1(y)$ we find that

$$(3.44) \qquad \hat{M}_1(\omega) = \int_{-1/2}^{1/2} e^{-i\omega y}\, dy = \frac{\sin(\omega/2)}{\omega/2};$$

then, using Plancherel's theorem for the Fourier transform of the convolution of two functions (see Papoulis (1962, p. 26)) we obtain

$$(3.45) \qquad \hat{M}_2(\omega) = \hat{M}_1(\omega)\hat{M}_1(\omega) = \left(\frac{\sin(\omega/2)}{\omega/2}\right)^2$$

and, recursively,

$$(3.46) \qquad \hat{M}_\mu(\omega) = \hat{M}_{\mu-1}(\omega)\hat{M}_1(\omega) = \left(\frac{\sin(\omega/2)}{\omega/2}\right)^\mu.$$

Likewise, through the simple change in scaling $x = yh$, we obtain

(3.47) $$\hat{\varphi}_0^q(\omega) = h\hat{M}_{q+1}(\omega h) = h\left(\frac{\sin(\omega h/2)}{\omega h/2}\right)^{q+1}.$$

We then find, by application of the sampling relation (1.79), that

(3.48) $$\hat{A}_1(\omega) = \frac{1}{h}\sum_{k=-\infty}^{\infty} \hat{\varphi}_0^{2q+1}\left(\omega + \frac{2k\pi}{h}\right)$$
$$= \left(\sin\left(\frac{\omega h}{2}\right)\right)^{2q+2} \sum_{k=-\infty}^{\infty} \frac{1}{(\omega h/2 + k\pi)^{2q+2}}$$

and

(3.49) $$\hat{A}_2(\omega) = -\frac{c}{h}\sum_{k=-\infty}^{\infty} i\left(\omega + \frac{2k\pi}{h}\right)\hat{\varphi}_0^{2q+1}\left(\omega + \frac{2k\pi}{h}\right)$$
$$= -ic\left(\sin\left(\frac{\omega h}{2}\right)\right)^{2q+2} \sum_{k=-\infty}^{\infty} \frac{(\omega + 2k\pi/h)}{(\omega h/2 + k\pi)^{2q+2}}.$$

Then, by (3.40)

(3.50) $$c^*(\omega) = \frac{c}{\omega} \frac{\sum_k \frac{(\omega + 2k\pi/h)}{(\omega h/2 + k\pi)^{2q+2}}}{\sum_k \frac{1}{(\omega h/2 + k\pi)^{2q+2}}}$$
$$= c\left[1 + \frac{2\pi}{\omega h} \frac{\sum_k \frac{k}{(\omega h/2 + k\pi)^{2q+2}}}{\sum_k \frac{1}{(\omega h/2 + k\pi)^{2q+2}}}\right].$$

It may be shown (Swartz and Wendroff (1974a)) that this results in

(3.51) $$c^* = c[1 + O((\omega h)^{2q+2})],$$

which is sometimes referred to as the "superconvergence" property of the B-spline Galerkin method. Note, however, that if instead of (3.38)–(3.39) a method of undetermined coefficients were used to obtain the $\{a_{1,k}\}$ and $\{a_{2,k}\}$ of $A_1\cdot$ and $A_2\cdot$, each with the same bandwidth $|k| \leq q$, then we would have $(4q+1)$ free coefficients, and the achievable rate of convergence of c^* would thus be

(3.52) $$c^* = c[1 + O((\omega h)^{4q})].$$

This is equal to (3.51) when $q = 1$, but is better when $q > 1$. Thus the "superconvergent" B-spline Galerkin algorithms are *not* giving the best rate of convergence that can be achieved for the given complexity, when $q > 1$.

It may also be shown that the second term in the brackets of (3.50) converges to zero when $q \to \infty$, provided $|\omega h| < \pi$ (the domain of convergence of the series (3.48) and (3.49)). In fact, the properties of this limit case are identical to those of the limit case ($K \to \infty$) of the explicit semi-discretizations examined in § 3.3.

The efficiency (meaning the accuracy achieved for a specified amount of computing work) of the B-spline Galerkin implicit semi-discretizations was compared with those of the finite difference explicit algorithms of § 3.1 by Swartz and Wendroff (1974a).

3.10. B-spline semi-discretizations of the wave equation. The same procedure, using B-splines with Galerkin's method to derive discrete operators, can be used to derive higher order *implicit* semi-discretizations of the wave equation

$$(3.53) \qquad \frac{\partial^2 U}{\partial t^2} = c^2 \frac{\partial^2 U}{\partial x^2}.$$

The solution is approximated by

$$(3.54) \qquad U(x, t) \simeq u(x, t) = \sum_n \varphi_n^q(x) v_n(t),$$

where the $\{v_n(t)\}$ are the coefficients of u in the basis $\{\varphi_n^q(x)\}$ of B-splines of degree q. The Galerkin equations are

$$(3.55) \qquad \begin{aligned} \frac{1}{h} \sum_m \langle \varphi_n^q, \varphi_m^q \rangle \frac{d^2 v_m}{dt^2} &= \frac{c^2}{h} \sum_m \left\langle \varphi_n^q, \frac{d^2 \varphi_m^q}{dx^2} \right\rangle v_m \\ &= -\frac{c^2}{h} \sum_m \left\langle \frac{d\varphi_n^q}{dx}, \frac{d\varphi_m^q}{dx} \right\rangle v_m, \end{aligned}$$

which may be written in operator notation as

$$(3.56) \qquad \mathbf{A}_1 \cdot \frac{d^2 v_n}{dt^2} = \mathbf{B}_2 \cdot v_n.$$

While the operators $\mathbf{A}_1 \cdot$ are the same as those defined in § 3.5, the $\mathbf{B}_2 \cdot$ are

$$(3.57) \qquad \mathbf{B}_2 \cdot = -\frac{c^2}{h} \sum_k \left\langle \frac{d\varphi_0^q}{dx}, \frac{d\varphi_k^q}{dx} \right\rangle \mathbf{E}^k \cdot .$$

As in § 3.7 this is also

$$(3.58) \qquad \mathbf{B}_2 \cdot = c^2 \sum_k \frac{d^2}{dx^2} \varphi_0^{2q+1}(kh) \mathbf{E}^k \cdot ,$$

which establishes the equivalence between a collocation method with splines of degree $2q+1$ and a Galerkin method with splines of degree q. As in the case of first order equations, identical evolution equations apply to:

(i) the coefficients $\{v_n(t)\}$ of a degree q B-spline Galerkin approximation of solutions of $U_{tt} = c^2 U_{xx}$;
(ii) the nodal values $\{u_n(t)\}$ of the same spline;
(iii) the coefficients $\{w_n(t)\}$ of a degree $(2q+1)$ B-spline collocation approximation;
(iv) the nodal values $\{u_n(t)\}$ of that spline.

In other words, the following equation holds:

$$\mathbf{A}_1 \cdot \frac{d^2 u_n}{dt^2} = \mathbf{B}_2 \cdot u_n, \tag{3.59}$$

where the $\{u_n(t)\}$ are as in (ii) or (iv).

A simple relation exists between the operators $\mathbf{A}_1 \cdot$ and $\mathbf{B}_2 \cdot$. We place the superscripts q in the notation $\mathbf{A}_1^q \cdot$, $\mathbf{B}_2^q \cdot$ to specify their order; thus

$$\mathbf{A}_1^q \cdot = \sum_k \varphi_0^{2q+1}(kh) \mathbf{E}^k \cdot, \tag{3.60}$$

$$\mathbf{B}_2^q \cdot = c^2 \sum_k \frac{d^2}{dx^2} \varphi_0^{2q+1}(kh) \mathbf{E}^k \cdot. \tag{3.61}$$

Then, from the identity (see, e.g., Schoenberg (1973, p. 12))

$$\frac{d}{dx} \varphi_n^q(x) = \frac{1}{h} \left[\varphi_n^{q-1}\left(x + \frac{h}{2}\right) - \varphi_n^{q-1}\left(x - \frac{h}{2}\right) \right],$$

we easily derive

$$\frac{d^2}{dx^2} \varphi_0^{2q+1}(x) = \frac{\mathbf{D}_2 \cdot}{h^2} \varphi_0^{2q-1}(x), \tag{3.62}$$

where $\mathbf{D}_2 \cdot$ is the central difference operator defined by (3.4). Thus, (3.61) is also:

$$\begin{aligned}\mathbf{B}_2^q \cdot &= \left(\frac{c}{h}\right)^2 \mathbf{D}_2 \cdot \sum_k \varphi_0^{2q-1}(kh) \mathbf{E}^k \cdot \\ &= \left(\frac{c}{h}\right)^2 \mathbf{D}_2 \cdot \mathbf{A}_1^{q-1} \cdot;\end{aligned} \tag{3.63}$$

(3.56) may be rewritten as

$$\mathbf{A}_1^q \cdot \frac{d^2 v_n}{dt^2} = \left(\frac{c}{h}\right)^2 \mathbf{D}_2 \cdot \mathbf{A}_1^{q-1} \cdot v_n \tag{3.64}$$

and (3.59) as

$$\mathbf{A}_1^q \cdot \frac{d^2 u_n}{dt^2} = \left(\frac{c}{h}\right)^2 \mathbf{D}_2 \cdot \mathbf{A}_1^{q-1} \cdot u_n. \tag{3.65}$$

3.11. Analysis. The algorithms defined by (3.56) are conservative. The approximation error is thus entirely described by the spurious dispersion due to the ω dependence of the numerical phase velocity. Proceeding as in § 2.9 we find

$$(3.66) \qquad c^*(\omega) = \frac{1}{\omega}\left(-\frac{\hat{B}_2^q(\omega)}{\hat{A}_1(\omega)}\right)^{1/2}$$

or, also

$$(3.67) \qquad c^*(\omega) = \frac{c}{\omega h}\left(-\frac{\hat{D}_2(\omega)\hat{A}_1^{q-1}(\omega)}{\hat{A}_1^q(\omega)}\right)^{1/2}$$

$$= \frac{c}{\omega h}\left(\frac{2(1-\cos(\omega h))\hat{A}_1^{q-1}(\omega)}{\hat{A}_1^q(\omega)}\right)^{1/2}$$

with

$$(3.68) \quad \begin{aligned} &\hat{A}_1^0(\omega) = 1, \\ &\hat{A}_1^1(\omega) = \tfrac{1}{3}[2+\cos(\omega h)], \\ &\hat{A}_1^2(\omega) = \tfrac{1}{60}[33+26\cos(\omega h)+\cos(2\omega h)], \\ &\hat{A}_1^3(\omega) = \tfrac{1}{2520}[1208+1191\cos(\omega h)+120\cos(2\omega h)+\cos(3\omega h)], \end{aligned}$$

and $\hat{D}_2(\omega)$ given by (3.7). See Fig. 3.5. The corresponding rates of convergence are easily derived by a series expansion. They are (Dougalis and Serbin (1981))

$q=1$. *Linear splines.*

$$c^* = c\left[1 + \frac{(\omega h)^2}{24} + O((\omega h)^4)\right].$$

$q=2$. *Quadratic splines.*

$$c^* = c\left[1 + \frac{(\omega h)^4}{1260} + O((\omega h)^6)\right].$$

$q=3$. *Cubic splines.*

$$c^* = c\left[1 + \frac{(\omega h)^6}{680400} + O(\omega h)^8)\right].$$

It may also be observed that these approximations "stiffen" the propagating medium, i.e., in all three cases

$$c^*(\omega) \geq c \quad \text{for } |\omega| \leq \frac{\pi}{h}.$$

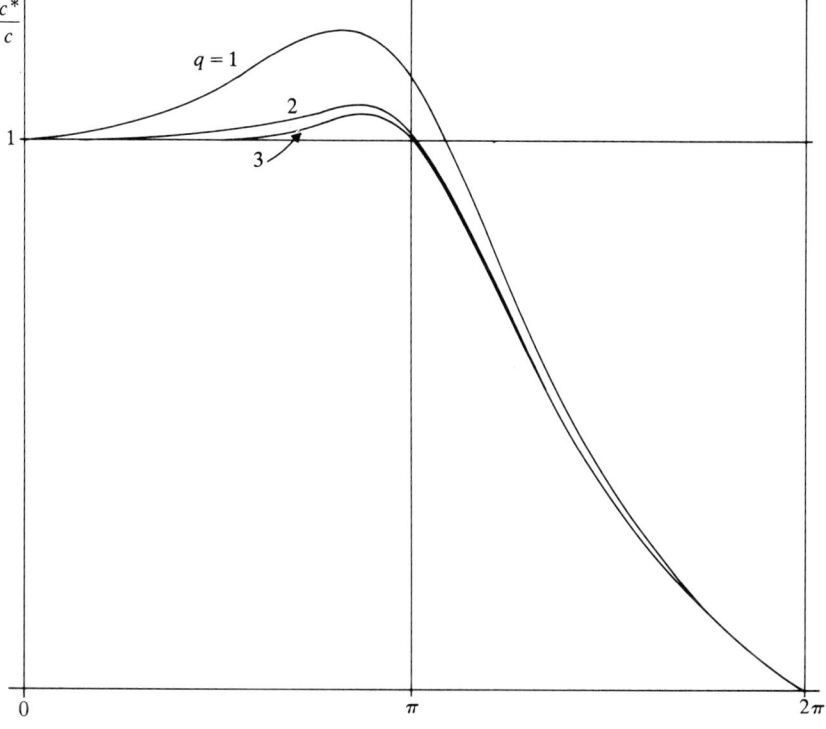

Fig. 3.5

Chapter 4

Full Discretizations

4.1. Fourier analysis. A Fourier analysis of full discretizations is obtained by seeking the time evolution of numerical solutions of the form

(4.1) $$u^j_{\omega,n} = v^j_\omega e^{i\omega x_n}.$$

As before, such solutions may be interpreted as "sinusoidal trial solutions" or, by considering ω to be a parameter with continuous domain $(-\pi/h, \pi/h)$, v^j_ω can be interpreted as the Fourier transform of numerical solutions. Inserting (4.1) into the discrete algorithm (1.48) gives of course simply

(4.2) $$u^{j+1}_{\omega,n} = M(\Delta t A \cdot, Z \cdot) u^j_{\omega,n}.$$

But, since

(4.3) $$A \cdot e^{i\omega x_n} = \hat{A}(\omega) e^{i\omega x_n},$$

this may also be expressed as

(4.4) $$v^{j+1}_\omega = M(\Delta t \hat{A}(\omega), Z \cdot) v^j_\omega,$$

which is a *scalar* time marching equation with parameter ω (by scalar we mean that the space or n dependence has been eliminated). To this equation we seek solutions of the form:

(4.5) $$\frac{v^{j+1}_\omega}{v^j_\omega} = z(\omega) = \text{constant}.$$

For such solutions to satisfy (4.4) we must have

(4.6) $$z(\omega) v^j_\omega = M(\Delta t \hat{A}(\omega), z(\omega)) v^j_\omega;$$

i.e., $z(\omega)$ must satisfy the characteristic equation

(4.7) $$z(\omega) - M(\Delta t \hat{A}(\omega), z(\omega)) = 0.$$

$z(\omega)$ is called an *amplification factor* for the discrete algorithm (4.2).

For one-step explicit methods, the expressions for z and $\mathbf{M}$ are identical. That the expressions for z and $\mathbf{M}$ are different in the other cases is illustrated in the following examples.

Euler's explicit method.

$$\mathbf{u}^{j+1} = \mathbf{u}^j + \Delta t \mathbf{A} \cdot \mathbf{u}^j,$$

(4.8a) $\qquad \mathbf{M} = 1 + \Delta t \mathbf{A} \cdot, \qquad z(\omega) = 1 + \Delta t \hat{A}(\omega).$

Euler's implicit method.

$$\mathbf{u}^{j+1} = \mathbf{u}^j + \Delta t \mathbf{A} \cdot \mathbf{u}^{j+1},$$

(4.8b) $\qquad \mathbf{M} = 1 + \Delta t \mathbf{A} \cdot \mathbf{Z} \cdot, \qquad z(\omega) = \dfrac{1}{1 - \Delta t \hat{A}(\omega)}.$

Leapfrog method.

$$\mathbf{u}^{j+1} = \mathbf{u}^{j-1} + 2\Delta t \mathbf{A} \cdot \mathbf{u}^j,$$

(4.8c) $\qquad \mathbf{M} = \mathbf{Z}^{-1} \cdot + 2\Delta t \mathbf{A} \cdot, \qquad z(\omega) = \Delta t \hat{A}(\omega) \pm \sqrt{(\Delta t \hat{A}(\omega))^2 + 1}.$

Crank–Nicolson method.

$$\mathbf{u}^{j+1} = \mathbf{u}^j + \frac{\Delta t}{2}(\mathbf{A} \cdot \mathbf{u}^{j+1} + \mathbf{A} \cdot \mathbf{u}^j),$$

(4.8d) $\qquad \mathbf{M} = 1 + \tfrac{1}{2}(\Delta t \mathbf{A} \cdot + \Delta t \mathbf{A} \cdot \mathbf{Z} \cdot), \qquad z(\omega) = \dfrac{1 + \Delta t \hat{A}(\omega)/2}{1 - \Delta t \hat{A}(\omega)/2}.$

Runge–Kutta methods.

$$\mathbf{u}^{j+1} = \sum_{k=0}^{K} \frac{(\Delta t \mathbf{A} \cdot)^k}{k!} \mathbf{u}^j,$$

(4.8e) $\qquad \mathbf{M} = \sum_{k=0}^{K} \dfrac{(\Delta t \mathbf{A} \cdot)^k}{k!}, \qquad z(\omega) = \sum_{k=0}^{K} \dfrac{(\Delta t \hat{A}(\omega))^k}{k!}.$

We may note that $z(\omega)$ is an approximation to $e^{\Delta t \hat{A}(\omega)}$, which is the expression for the ratio $v_\omega^{j+1}/v_\omega^j$ if the semi-discretization were integrated exactly.

Approximations of e^x by a rational function (i.e., the ratio of two polynomials) are known as Padé approximants (Table 4.1).

The amplification factor for both Euler's methods, (4.8a) and (4.8b), are first order Padé approximants to $e^{\Delta t \hat{A}(\omega)}$, and its expression for the Crank–Nicolson marching method is a second order approximant. Padé approximants of the $(n, 0)$ type (those in the top line of Table 4.1) correspond to the classical explicit Runge–Kutta family of methods. (Other aspects of the relation between time marching methods and the approximation of e^x are discussed in Rosenbrock (1963).)

TABLE 4.1
Padé approximants to e^x

	$m=0$	$m=1$	$m=2$	$m=3$
$n=0$	$\dfrac{1}{1}$	$\dfrac{1+x}{1}$	$\dfrac{1+x+\frac{1}{2}x^2}{1}$	$\dfrac{1+x+\frac{1}{2}x^2+\frac{1}{6}x^3}{1}$
$n=1$	$\dfrac{1}{1-x}$	$\dfrac{1+\frac{1}{2}x}{1-\frac{1}{2}x}$	$\dfrac{1+\frac{2}{3}x+\frac{1}{6}x^2}{1-\frac{1}{3}x}$	$\dfrac{1+\frac{3}{4}x+\frac{1}{4}x^2+\frac{1}{24}x^3}{1-\frac{1}{4}x}$
$n=2$	$\dfrac{1}{1-x+\frac{1}{2}x^2}$	$\dfrac{1+\frac{1}{3}x}{1-\frac{2}{3}x+\frac{1}{6}x^2}$	$\dfrac{1+\frac{1}{2}x+\frac{1}{12}x^2}{1-\frac{1}{2}x+\frac{1}{12}x^2}$	$\dfrac{1+\frac{3}{5}x+\frac{3}{20}x^2+\frac{1}{60}x^3}{1-\frac{2}{5}x+\frac{1}{20}x^2}$
$n=3$	$\dfrac{1}{1-x+\frac{1}{2}x^2-\frac{1}{6}x^3}$	$\dfrac{1+\frac{1}{4}x}{1-\frac{3}{4}x+\frac{1}{4}x^2-\frac{1}{24}x^3}$	$\dfrac{1+\frac{2}{5}x+\frac{1}{20}x^2}{1-\frac{3}{5}x+\frac{3}{20}x^2-\frac{1}{60}x^3}$	$\dfrac{1+\frac{1}{2}x+\frac{1}{10}x^2+\frac{1}{120}x^3}{1-\frac{1}{2}x+\frac{1}{10}x^2-\frac{1}{120}x^3}$
$n=4$	$\dfrac{1}{1-x+\frac{1}{2}x^2-\frac{1}{6}x^3+\frac{1}{24}x^4}$	$\dfrac{1+\frac{1}{5}x}{1-\frac{4}{5}x+\frac{3}{16}x^2-\frac{1}{15}x^3+\frac{1}{120}x^4}$	$\dfrac{1+\frac{1}{3}x+\frac{1}{30}x^2}{1-\frac{2}{3}x+\frac{1}{5}x^2-\frac{1}{30}x^3+\frac{1}{360}x^4}$	$\dfrac{1+\frac{3}{7}x+\frac{1}{14}x^2+\frac{1}{210}x^3}{1-\frac{4}{7}x+\frac{1}{7}x^2-\frac{2}{210}x^3+\frac{1}{840}x^4}$

4.2. Stability. The von Neumann method of numerical stability analysis[1] is a Fourier analysis. It consists in following the time evolution of numerical sinusoidal solutions of the form (4.1), with the condition that the absolute value of the corresponding amplification factors (4.5) should not exceed unity for numerical stability.

The analysis is simplified in practice by defining "stability regions" in the complex plane.[2] For a given marching method characterized by an operator M, we consider the characteristic equation (4.7) as a mapping between the complex planes $\Delta t \hat{A}(\omega)$ and $z(\omega)$. To the stable unit disk in the latter

(4.9) $$|z(\omega)| \leq 1$$

there corresponds a region S in the former which is the *stability region* for that time marching method.

For example, for Euler's explicit method, the characteristic equation is

(4.10) $$z(\omega) - 1 - \Delta t \hat{A}(\omega) = 0$$

and the stability region illustrated in Fig. 4.1 corresponds to (4.9).[3]

For a fully discrete algorithm using M to be stable, it is necessary that $\Delta t \hat{A}(\omega)$ of the semi-discretization be contained in S for all values of ω. For example, to the central differences semi-discretization

(4.11) $$\frac{du_n}{dt} = -c\left(\frac{u_{n+1} - u_{n-1}}{2h}\right),$$

there corresponds

(4.12) $$\Delta t \hat{A}(\omega) = -i\left(\frac{c\Delta t}{h}\right) \sin(\omega h)$$

whose locus is the segment of the imaginary axis $[-i(c\Delta t/h), i(c\Delta t/h)]$. Marching (4.11) in time with Euler's explicit method is thus unstable for all values of the Courant number $(c\Delta t/h)$.

On the other hand, the leapfrog method (4.8c) has a stability region S which consists of the segment $[-i, i]$ of the imaginary axis in the $\Delta t \hat{A}$-plane (Fig. 4.2b). Using this method to march (4.11) in time results in a numerically stable algorithm if

(4.13) $$\left|\frac{c\Delta t}{h}\right| \leq 1.$$

---

[1] Von Neumann and Richtmyer (1950). See also O'Brien, Hyman and Kaplan (1950).
[2] Vichnevetsky (1972a).
[3] Stability regions are symmetric with respect to the real axis. It is thus sufficient to graph the upper half only, as is done in Fig. 4.2.

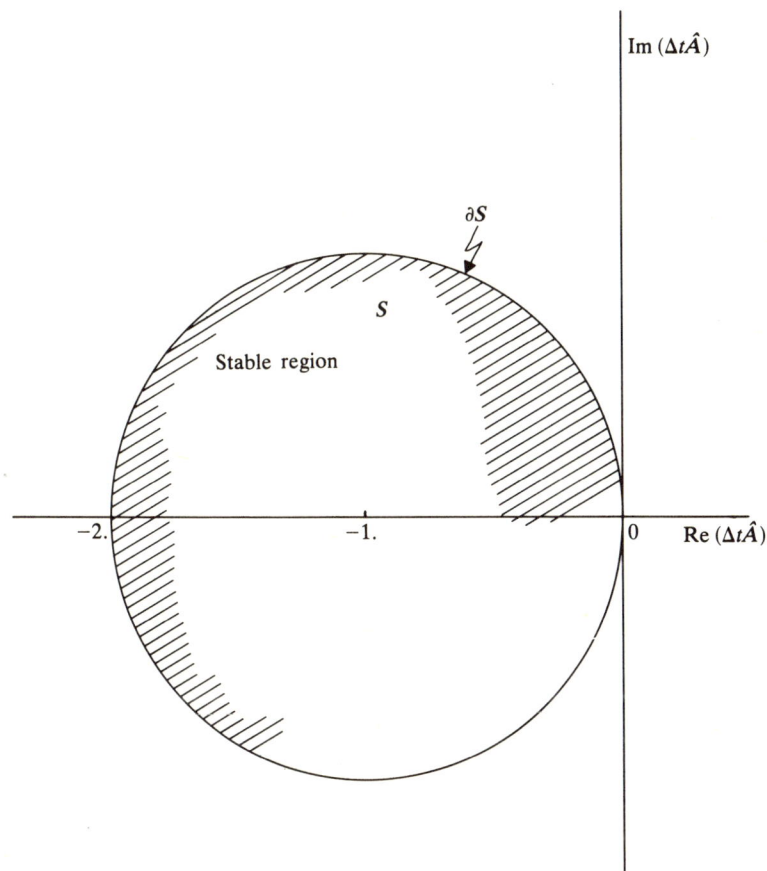

FIG. 4.1. *Stability region for Euler's explicit method.*

This is the historical Courant–Friedrichs–Lewy condition, which was derived in their 1928 paper.

Stability regions for several common marching methods are given in Fig. 4.2.

The condition of numerical stability for these methods when applied to the hyperbolic semi-discretization (4.11) may be expressed as

$$\left|\frac{c\Delta t}{h}\right| \leq S_I, \tag{4.14}$$

where S_I is the value of $|\Delta t \hat{A}|$ at which the stability boundary ∂S intersects the imaginary axis.

The values of S_I for the time marching methods shown in Figs. 4.2a–d are summarized in Table 4.2.

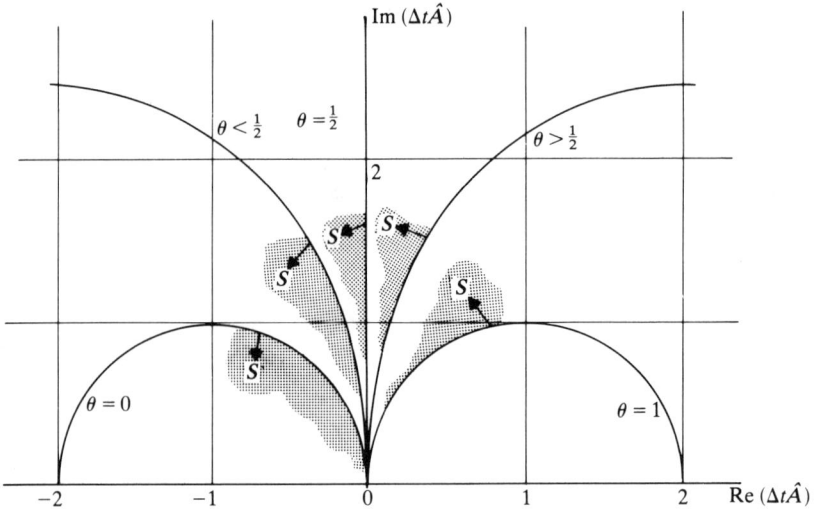

FIG. 4.2a. *Stability regions for the general implicit method,* $\mathbf{u}^{j+1} = \mathbf{u}^j + \Delta t(\theta \mathbf{A} \cdot \mathbf{u}^{j+1} + (1-\theta)\mathbf{A} \cdot \mathbf{u}^j)$.

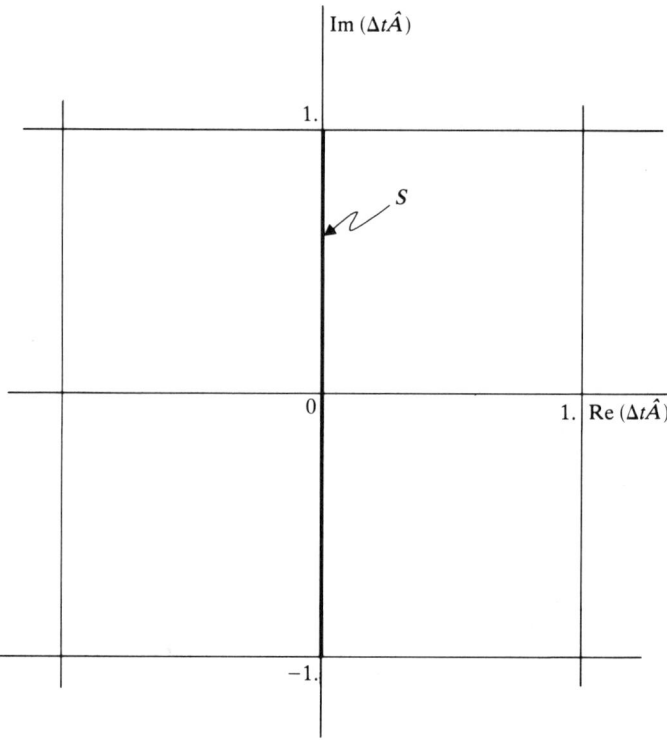

FIG. 4.2b. *Stability region for the leapfrog method,* $\mathbf{u}^{j+1} = \mathbf{u}^{j-1} + 2\Delta t \mathbf{A} \cdot \mathbf{u}^j$.

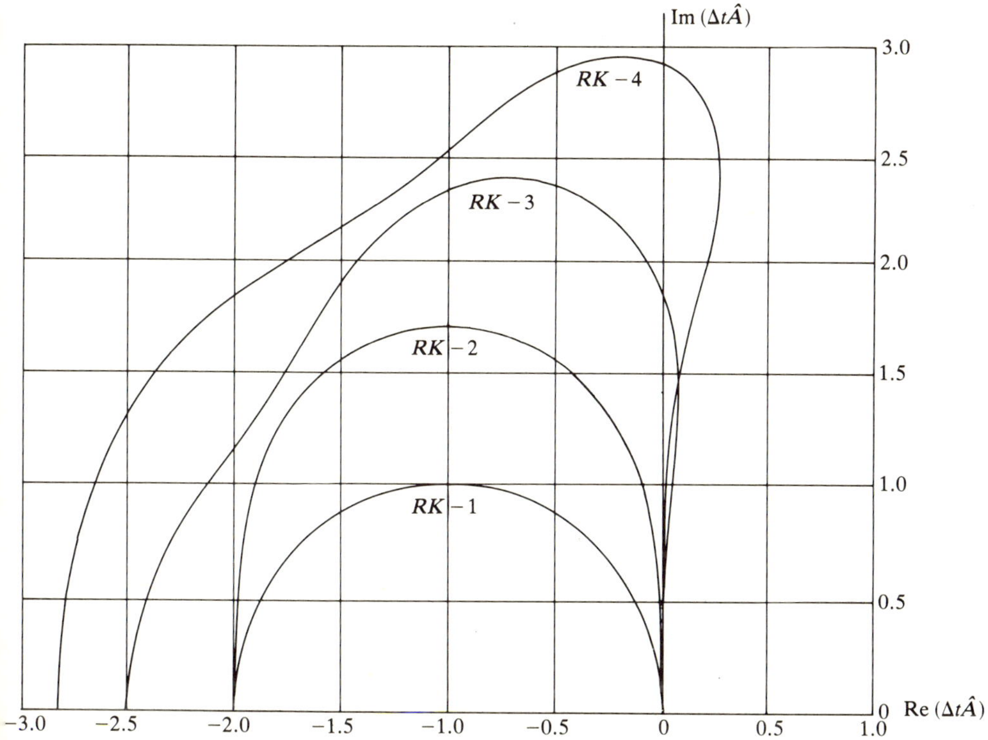

FIG. 4.2c. *Stability regions for the Runge–Kutta methods of order 1, 2, 3 and 4.*

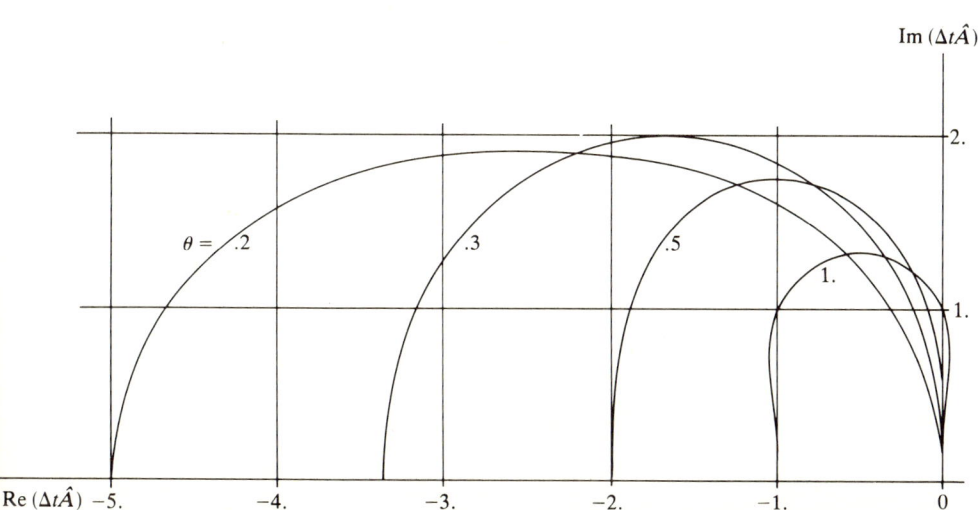

FIG. 4.2d. *Stability regions for the general explicit method (Vichnevetsky (1979))*,
$$\mathbf{u}^{j+\theta} = \mathbf{u}^j + \theta \Delta t \mathbf{A} \cdot \mathbf{u}^j,$$
$$\mathbf{u}^{j+1} = \mathbf{u}^j + \Delta t \mathbf{A} \cdot \mathbf{u}^{j+\theta}.$$

TABLE 4.2

Numerical stability conditions in the integration of the hyperbolic semi-discretization $du_n/dt = -c/2h(u_{n+1} - u_{n-1})$ by various discrete time marching methods

Time marching method (number indicates equation in the text)	S_I	Stability condition $\left\|\dfrac{c\Delta t}{h}\right\| \leq S_I$
Euler's explicit (4.8a)	0	always unstable
Euler's implicit (4.8b)	∞	always stable
Leapfrog (4.8c)	1	$\left\|\dfrac{c\Delta t}{h}\right\| \leq 1$
General implicit (Fig. 4.2a)	∞ for $\theta \geq \frac{1}{2}$ 0 for $\theta < \frac{1}{2}$	stable for $\theta \geq \frac{1}{2}$ unstable for $\theta < \frac{1}{2}$
General explicit (Vichnevetsky) (Fig. 4.2d)	$\left(\dfrac{\sqrt{2\theta-1}}{\theta}\right)$ for $\theta > \frac{1}{2}$ 0 for $\theta \leq \frac{1}{2}$	$\left\|\dfrac{c\Delta t}{h}\right\| \leq \left(\dfrac{\sqrt{2\theta-1}}{\theta}\right)$ for $\theta > \frac{1}{2}$ unstable for $\theta \leq \frac{1}{2}$
Runge–Kutta second order	0	always unstable
Runge–Kutta third order	1.8	$\left\|\dfrac{c\Delta t}{h}\right\| \leq 1.8$
Runge–Kutta fourth order	2.85	$\left\|\dfrac{c\Delta t}{h}\right\| \leq 2.85$

4.3. Velocity and amplitude error. The true ratio for sinusoidal solutions

$$(4.15) \quad \frac{U_\omega(x, t+\Delta t)}{U_\omega(x, t)} = e^{-i\omega c \Delta t}$$

is replaced in the numerical approximations by

$$(4.16) \quad \frac{u_{\omega,n}^{j+1}}{u_{\omega,n}^{j}} = z(\omega),$$

where $z(\omega)$ is the amplification factor of the scheme. Comparing these two expressions and separating amplitude $|z(\omega)|$ and phase $\angle z(\omega)$ leads us to write

$$(4.17) \quad z(\omega) = |z(\omega)| e^{i\angle z(\omega)} = |z(\omega)| e^{-i\omega c^*(\omega)\Delta t}.$$

This defines the phase velocity of a full discretization implicitly as

$$(4.18) \quad c^*(\omega) = \frac{-\angle z(\omega)}{\omega \Delta t}.$$

Since the amplitude of (4.15) is 1, the difference

(4.19) $$|z(\omega)| - 1$$

represents an error. We define

(4.20) $$\varepsilon_A = \frac{|z(\omega)| - 1}{\Delta t}$$

as the *amplitude error* of the approximation. Since for all consistent time marching methods we have

(4.21) $$z(\omega) = 1 + \Delta t \hat{A}(\omega) + \text{higher order terms},$$

we find

(4.22) $$\lim_{\Delta t \to 0} \varepsilon_A = \lim_{\Delta t \to 0} \frac{|1 + \Delta t \hat{A}(\omega)| - 1}{\Delta t} = \text{Re } \hat{A}(\omega).$$

When $\Delta t \to 0$, the amplitude error tends to that due to the spatial semi-discretization alone, as expected. Likewise, (4.18) becomes

(4.23) $$\lim_{\Delta t \to 0} c^*(\omega) = \lim_{\Delta t \to 0} \frac{-\angle(1 + \Delta t \hat{A}(\omega))}{\omega \Delta t} = -\frac{\text{Im } \hat{A}(\omega)}{\omega},$$

which is precisely the numerical velocity of the spatial semi-discretization.

4.4. Examples. We limit our analysis of examples to a few full discretizations which are conservative, that is, $|z(\omega)| = 1$ for all ω and the only error is spurious dispersion ($c^*(\omega) \neq c$).

The first two cases considered start from the explicit 3-point semi-discretization

$$\frac{du_n}{dt} = -c \left(\frac{u_{n+1} - u_{n-1}}{2h} \right) \equiv A \cdot u_n$$

with integration in time approximated by the leapfrog and Crank–Nicolson methods respectively.

The third case starts from the 2-point implicit semi-discretization

$$\frac{1}{2} \left(\frac{du_n}{dt} + \frac{du_{n+1}}{dt} \right) = -c \left(\frac{u_{n+1} - u_n}{h} \right)$$

with time marching also by the Crank–Nicolson method. (This is the "box" method of approximation, see Table 2.1.)

The expressions for c^* are summarized in Table 4.3, and the corresponding ratios c^*/c are illustrated in Fig. 4.3. Note that for each method c^* is now not only a function of ωh, but also of the Courant number and hence has a dependence on Δt.

TABLE 4.3

Algorithm	Numerical phase velocity ($R \equiv c\Delta t/h$ = Courant number)
Central differences/leapfrog $$u_n^{j+1} = u_n^{j-1} + 2\Delta t \mathbf{A} \cdot u_n^j$$ $$\mathbf{A} \cdot = -c\left(\frac{\mathbf{E} - \mathbf{E}^{-1}}{2h}\right) \cdot$$	$$\frac{c^*}{c} = \frac{\sin^{-1}(R\sin(\omega h))}{R\omega h}$$
Central differences/Crank–Nicolson $$u_n^{j+1} = u_n^j + \frac{\Delta t}{2}(\mathbf{A} \cdot u_n^{j+1} + \mathbf{A} \cdot u_n^j)$$ $$\mathbf{A} \cdot = -c\left(\frac{\mathbf{E} - \mathbf{E}^{-1}}{2h}\right) \cdot$$	$$\frac{c^*}{c} = \frac{2\tan^{-1}((R/2)\sin(\omega h))}{R\omega h}$$
Box method $$u_n^{j+1} = u_n^j + \frac{\Delta t}{2}(\mathbf{A} \cdot u_n^{j+1} + \mathbf{A} \cdot u_n^j)$$ $$\mathbf{A} \cdot = -c\left(\frac{\mathbf{E}+1}{2}\right)^{-1} \cdot \left(\frac{\mathbf{E}-1}{h}\right) \cdot$$	$$\frac{c^*}{c} = \frac{2\tan^{-1}(R\tan(\omega h/2))}{R\omega h}$$

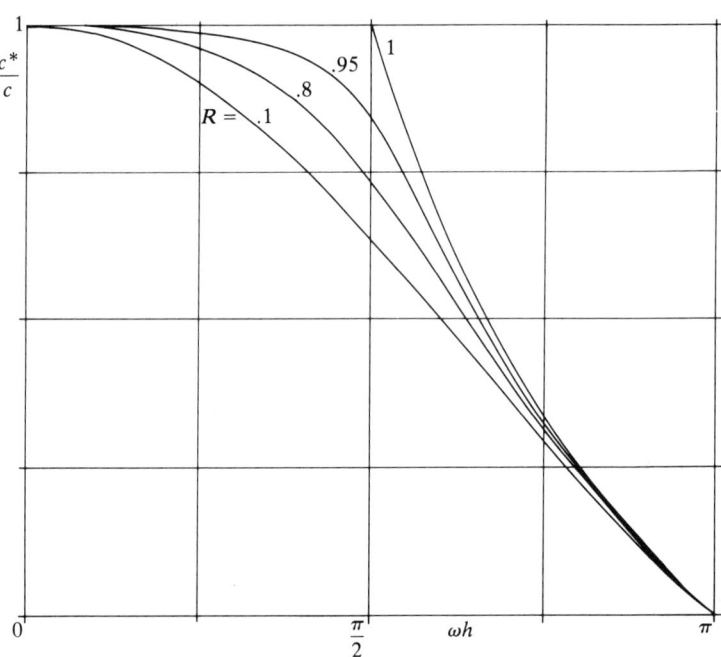

FIG. 4.3a. *Numerical phase velocity of the central differences/leapfrog method. Note the accuracy when $R = 1$ and $\omega h \leq \pi/2$. When $R \to 0$, the error tends to that due to the spatial semi-discretization alone.*

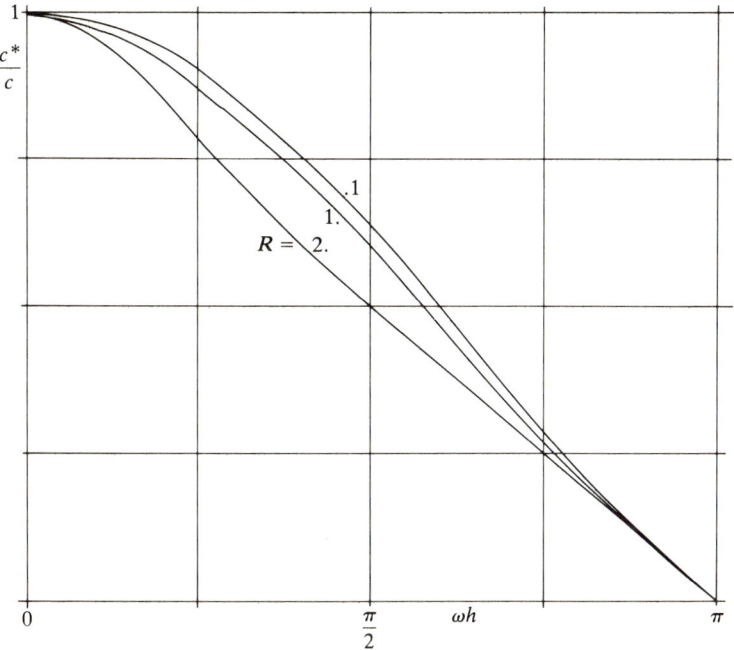

FIG. 4.3b. *Numerical phase velocity of the central differences/Crank–Nicolson method. Errors due to the approximation in space and time directions are of the same sign, and are therefore additive (they were of opposite sign and to a degree cancelled each other in the case illustrated in Fig. 4.3a).*

4.5. Time marching methods for second order equations. Analytic solutions of the second order equation

$$\text{(4.24)} \qquad \frac{d^2 y}{dt^2} + \lambda^2 y = 0$$

satisfy

$$\text{(4.25)} \qquad \frac{y(t+\Delta t) + y(t-\Delta t)}{2y(t)} = \cos(\lambda \Delta t).$$

Thus, rational approximations of $\cos(\lambda \Delta t)$ will lead to the development of discrete time marching methods in the same sense that the Padé rational approximants of e^x did with first order equations (§ 4.1). This is the basis of the "cosine methods" developed by Baker, Dougalis and Serbin (1981). For instance, the approximation

$$\text{(4.26)} \qquad \cos(\lambda \Delta t) = 1 - \frac{(\lambda \Delta t)^2}{2} + O(\Delta t^4)$$

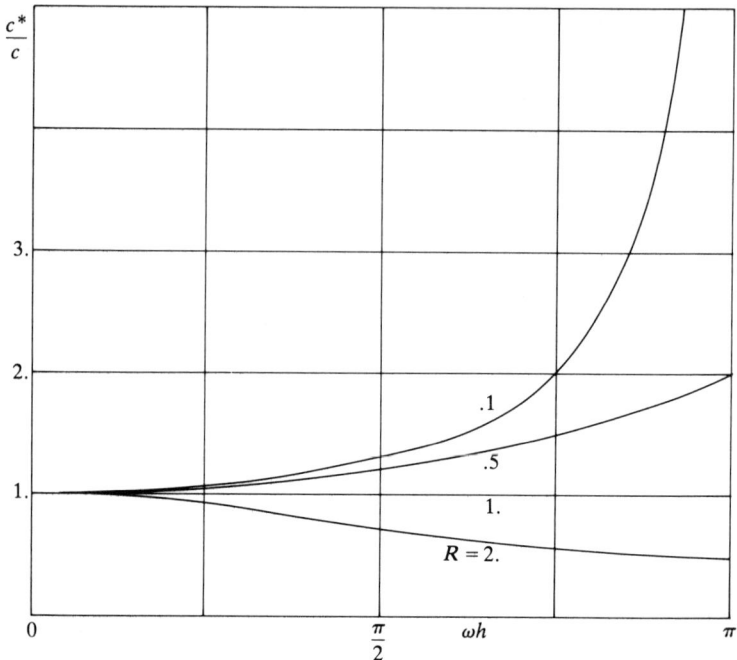

FIG. 4.3c. *Numerical phase velocity for the box method. For $\omega h = \pi$, we have $c^*/c = 1/R$.*

becomes

(4.27) $$y^{j+1} - 2y^j + y^{j-1} = -(\lambda \Delta t)^2 y^j,$$

which is the simple explicit method used in (2.48). To the rational approximation

(4.28) $$\cos(\lambda \Delta t) = \frac{1 - \frac{5}{12}(\lambda \Delta t)^2}{1 + \frac{1}{12}(\lambda \Delta t)^2} + O(\Delta t^6)$$

corresponds the Störmer–Numerov implicit algorithm (compare with (2.60))

(4.29) $$y^{j+1} - 2y^j + y^{j-1} = -\frac{(\lambda \Delta t)^2}{12}(y^{j+1} + 10y^j + y^{j-1}).$$

Other approximations based on the cosine method which are suitable for the time discretization of second order hyperbolic equations are given in Dougalis and Serbin (1981).

Chapter 5

Damping, Diffusion and Filtering

5.1. Spurious diffusion. Amplitude decay of sinusoidal components of numerical solutions of hyperbolic equations is a byproduct of the approximation for those methods which are "nonconservative". On the other hand, amplitude decay also exists with solutions of hyperbolic equations to which a diffusion term has been added; the simplest model is the advection equation with diffusion

(5.1) $$\frac{\partial U}{\partial t} + c\frac{\partial U}{\partial x} = \sigma\frac{\partial^2 U}{\partial x^2}.$$

From the strict mathematical viewpoint this equation is of course no longer hyperbolic, but parabolic. However, with σ small, one can say that the equation is still "almost hyperbolic". The Fourier transform of solutions of this equation is

(5.2) $$\hat{U}(\omega, t) = \hat{U}(\omega, 0)\, e^{-ic\omega t}\, e^{-\sigma\omega^2 t},$$

whence amplitude decay with time as

(5.3) $$|\hat{U}(\omega, t)| = |\hat{U}(\omega, 0)|\, e^{-\sigma\omega^2 t}.$$

Conversely, when a numerical method for the pure advection equation

(5.4) $$\frac{\partial U}{\partial t} + c\frac{\partial U}{\partial x} = 0$$

produces an amplitude decay which depends on ω as

(5.5) $$|\hat{u}(\omega, t)| = |\hat{u}(\omega, 0)|\, e^{-\gamma\omega^2 t}$$

then we may equate γ to a *diffusion constant* which is not present in the original equation, but which is a *spurious* result of the approximation; i.e., we shall say that the numerical algorithm is an approximation of

(5.6) $$\frac{\partial u}{\partial t} + c\frac{\partial u}{\partial x} = \gamma\frac{\partial^2 u}{\partial x^2}.$$

It turns out that the ω^2-dependence displayed in (5.5) occurs, when $\omega \to 0$, for all methods with *first order degree of accuracy*. Methods with a higher degree of accuracy have an amplitude decay which may depend on a power of ω higher than two. Such decay cannot be "modeled" by a spurious diffusion term, but rather by a term of the form $k\, \partial^p u/\partial x^p$, where p is larger than 2 (or amplitude decay may of course be nonexistent).

For semi-discrete approximations, spurious diffusion may only occur if $\operatorname{Re} \hat{A}(\omega) \neq 0$, i.e.,

(5.7) $$\operatorname{Re} \hat{A}(\omega) = -\gamma \omega^2 + \text{higher order terms}.$$

For full discretizations, spurious diffusion is related to the amplification factor by the relation

(5.8) $$|z(\omega)| = e^{-\gamma \omega^2 \Delta t} + \text{higher order terms}.$$

In both cases, γ is the coefficient of the spurious diffusion term, as shown in (5.6).

Example 5.1. As a first example, consider the 2-point semi-discretization of the advection equation

(5.9) $$\frac{du_n}{dt} = -c\left(\frac{u_n - u_{n-1}}{h}\right) \equiv \boldsymbol{A} \cdot u_n$$

(sometimes called the "backward" or "upwind" difference approximation). The spectral function is

(5.10) $$\hat{A}(\omega) = -c\left(\frac{1 - e^{-i\omega h}}{h}\right) = -\frac{c}{h}\left(i \sin(\omega h) + 2 \sin^2\left(\frac{\omega h}{2}\right)\right),$$

whence

(5.11) $$\begin{aligned} \bar{u}(\omega, t) &= \bar{u}(\omega, 0)\, e^{\hat{A}(\omega) t} \\ &= \bar{u}(\omega, 0)\, e^{-i\omega(c + O(h^2))t}\, e^{(-c\omega^2(h/2) + O(h^3))t} \end{aligned}$$

which leads to

(5.12) $$|\bar{u}(\omega, t)| = |\bar{u}(\omega, 0)|\, e^{-\gamma \omega^2 t}\, (1 + \text{higher order terms})$$

with γ given by the formula

(5.13) $$\gamma = \frac{ch}{2}.$$

Example 5.2. The general implicit time marching method

(5.14) $$u_n^{j+1} = u_n^j + \Delta t(\theta \boldsymbol{A} \cdot u_n^{j+1} + (1 - \theta)\boldsymbol{A} \cdot u_n^j)$$

has first order accuracy when θ is taken different from $\frac{1}{2}$. To study the effect of the time marching error alone, we let $\hat{A}(\omega) = -ic\omega$. Then the amplification factor $z(\omega)$ becomes

$$\text{(5.15)} \quad z(\omega) = \frac{1+(1-\theta)(-ic\omega\Delta t)}{1-\theta(-ic\omega\Delta t)}.$$

Expanding in powers of Δt, we find

$$z(\omega) = e^{-i\omega(c\Delta t + O(\Delta t^2))} e^{-\gamma\omega^2 \Delta t + O(\Delta t^3)},$$

where

$$\text{(5.16)} \quad \gamma = c^2 \Delta t (\theta - \tfrac{1}{2})$$

is the spurious diffusion constant introduced by the approximation.

5.2. Limitations of the spurious diffusion model. When expressed in the frequency domain, the validity of the concept of "spurious diffusion" of an approximation is restricted to small values of ωh and may fail to describe what happens for larger values. This is true in particular for sinusoidal components of wavelength $\lambda = 2h$ (or frequency $\omega = \pi/h$).

Consider, for example, the simple central difference semi-discretization

$$\text{(5.17)} \quad \frac{du_n}{dt} = -c\left(\frac{u_{n+1}-u_{n-1}}{2h}\right) \equiv \mathbf{A} \cdot u_n.$$

We have in the frequency domain

$$\text{(5.18)} \quad \hat{A}\left(\frac{\pi}{h}\right) = -ic\frac{\sin(\pi)}{h} = 0.$$

Incidentally, this says that numerical components of wavelength $2h$ have a time evolution equation identical to that of components of infinite wavelength ($\omega h = 0$). Consider now the time marching of (5.17) with the general implicit method (5.14). We find that

$$\text{(5.19)} \quad z(\omega) = \frac{1-ic(1-\theta)\Delta t \sin(\omega h)/h}{1+ic\theta \Delta t \sin(\omega h)/h},$$

and hence

$$\text{(5.20)} \quad z\left(\frac{\pi}{h}\right) = 1.$$

Thus, *there is no damping of the $2h$ wavelength components* for any value of θ.

The damping of sinusoidal components as a function of ωh in $[0, \pi]$ is illustrated in Fig. 5.1.

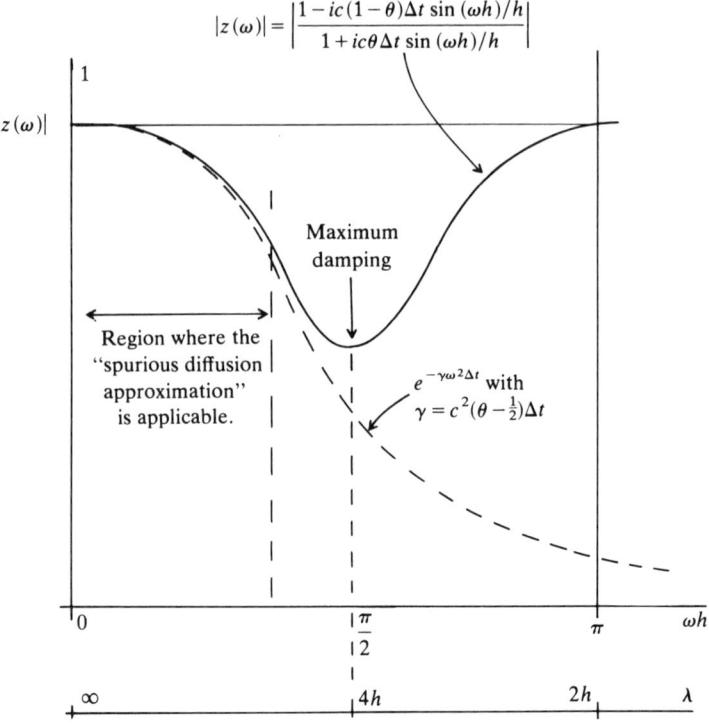

FIG. 5.1. *Damping of sinusoidal components in the numerical solution to $U_t + cU_x = 0$ approximated in space with the simple 3-point explicit central differences semi-discretization and in time by the general implicit marching method.*

The fact that no numerically induced damping occurs for sinusoidal solutions of wavelength $2h$ will remain true for all methods that use the semi-discretization (5.17) as a starting point. When desired, there are other ways in which short wavelength components of the solution may be damped. A possible method which applies to second order hyperbolic systems is described in § 5.3 below. Another way is to use special full discretizations for hyperbolic systems that contain what amounts to a damping operator, of which the Lax Wendroff method is the archetype (Lax and Wendroff (1960)). Finally, one may simply use numerical filtering, as described in § 5.4 and following.

5.3. Spurious diffusion in discretizations of the wave equation. Consider the wave equation

(5.21) $$\frac{\partial^2 U}{\partial t^2} = c^2 \frac{\partial^2 U}{\partial x^2},$$

DAMPING, DIFFUSION AND FILTERING

and the two equivalent first order systems into which it can be transformed

(5.22a) $$\frac{\partial U}{\partial t} = c\frac{\partial V}{\partial x}, \quad \frac{\partial V}{\partial t} = c\frac{\partial U}{\partial x},$$

(5.22b) $$\frac{\partial U}{\partial t} = V, \quad \frac{\partial V}{\partial t} = c^2\frac{\partial^2 U}{\partial x^2}.$$

After semi-discretization, (5.22a) becomes

(5.23) $$\frac{d}{dt}\begin{pmatrix}u_n\\v_n\end{pmatrix} = c\left(\frac{\boldsymbol{E}-\boldsymbol{E}^{-1}}{2h}\right) \cdot \begin{pmatrix}0 & 1\\1 & 0\end{pmatrix} \cdot \begin{pmatrix}u_n\\v_n\end{pmatrix},$$

which is equivalent, modulo a linear transformation, to two scalar equations of the form

(5.24) $$\frac{dw_n}{dt} = \pm c\left(\frac{\boldsymbol{E}-\boldsymbol{E}^{-1}}{2h}\right) \cdot w_n.$$

The spurious damping associated with the time discretization of (5.22a) with an implicit method is thus that described by (5.19). In particular, components of wavelength $2h$ are unaffected.

The situation is different with (5.22b), which leads to the semi-discrete equations

(5.25) $$\frac{d}{dt}\begin{pmatrix}u_n\\v_n\end{pmatrix} = \begin{pmatrix}0 & 1\\ \frac{c^2}{h^2}(\boldsymbol{E}-2+\boldsymbol{E}^{-1}) & 0\end{pmatrix} \cdot \begin{pmatrix}u_n\\v_n\end{pmatrix}.$$

For sinusoidal solutions, this system becomes

(5.26) $$\frac{d}{dt}\begin{pmatrix}\bar{u}(\omega,t)\\\bar{v}(\omega,t)\end{pmatrix}e^{i\omega x_n} = \begin{pmatrix}0 & 1\\ -\frac{4c^2}{h^2}\sin^2\left(\frac{\omega h}{2}\right) & 0\end{pmatrix} \cdot \begin{pmatrix}\bar{u}(\omega,t)\\\bar{v}(\omega,t)\end{pmatrix}e^{i\omega x_n},$$

which may be transformed by diagonalization into two linear systems of the form

(5.27) $$\frac{d}{dt}\bar{w}(\omega,t) = \pm\frac{2ic}{h}\sin\left(\frac{\omega h}{2}\right)\bar{w}(\omega,t).$$

The amplification factors associated with the general implicit time discretization are

(5.28) $$z(\omega) = \frac{1 \mp 2ic(1-\theta)\Delta t \sin(\omega h/2)/h}{1 \pm 2ic\theta\Delta t \sin(\omega h/2)/h}.$$

For $\omega h \to 0$, the spurious diffusion constant is as for (5.15):

(5.29)
$$\gamma = c^2 \Delta t (\theta - \tfrac{1}{2}).$$

The important difference is that here $|z(\omega)|$ attains its *minimum* for $\omega h = \pi$, i.e., for the components of wavelength $2h$ (see Fig. 5.2).

There are applications where either formulation may be used and where damping out the $2h$ oscillations is indeed desirable (those oscillations are

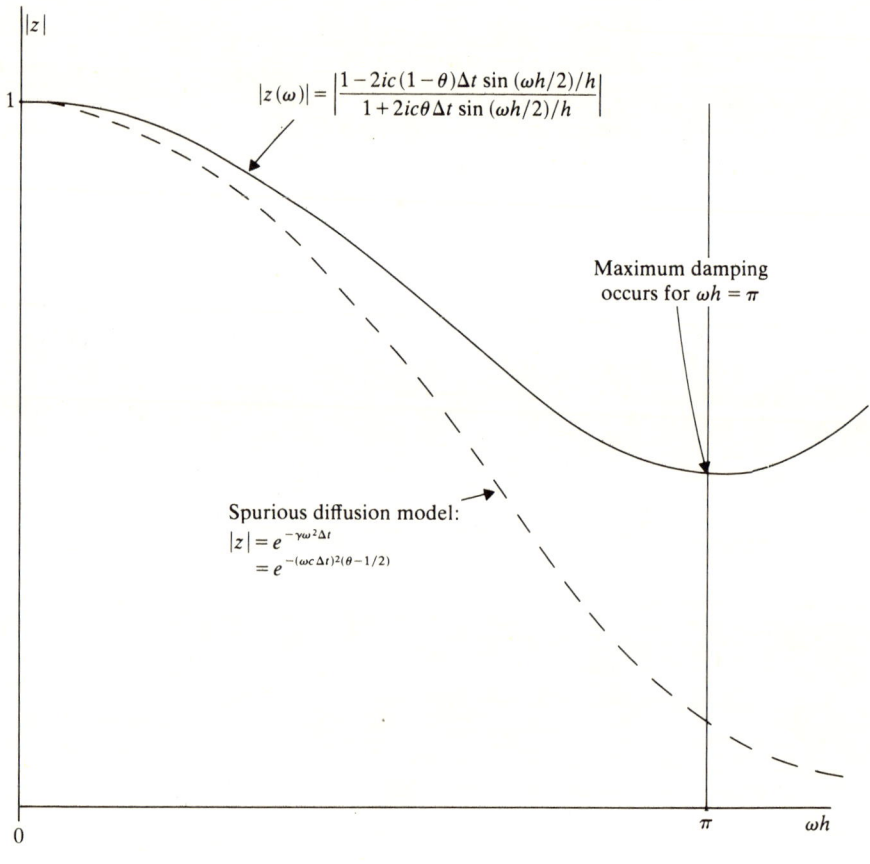

FIG. 5.2. *Damping of sinusoidal components in numerical solutions of $U_{tt} - c^2 U_{xx} = 0$, discretized in space with 3-point central differences and in time with the general implicit method. Maximum damping occurs in this case for $\lambda = 2h$.*

always spurious, see in particular Chapt. 7). An example is that of the hydrologists' shallow wave equation. In its normal form it is a system of two first order equations akin to (5.22a), and numerical simulations based on this formulation often display an excessive amount of unwanted oscillation. As could have been expected, one then finds in the relevant literature mention of the possible advantages of the other formulation. For example, Gray (1980) reports the development of "... a wave equation model (of the shallow wave equation) ... which has suppressed the short wavelength noise in a number of test simulations."

5.4. Filtering. The operation of modifying a continuous function selectively in the frequency domain is known by engineers as "filtering". A well-known example of filtering is that used in audio amplifiers to selectively increase or decrease the amplitude of sound in a given frequency band. *Numerical filtering* may be defined as follows:

Given the discrete set

(5.30) $$\{u_n; n = \cdots, -1, 0, 1, 2, \cdots\},$$

which has the Fourier transform $\bar{u}(\omega)$, the application of a discrete operator

(5.31) $$\boldsymbol{F} \cdot \equiv \sum_{k=-N_1}^{N_2} f_k \boldsymbol{E}^k \cdot$$

results in the new "filtered" set $\{u_n^F\}$ given by

(5.32) $$u_n^F = \sum_{k=-N_1}^{N_2} f_k u_{n+k}.$$

Its discrete Fourier transform is

(5.33) $$\bar{u}^F(\omega) = \hat{F}(\omega)\bar{u}(\omega),$$

where

(5.34) $$\hat{F}(\omega) = \frac{\boldsymbol{F} \cdot e^{i\omega x_n}}{e^{i\omega x_n}} = \sum_k f_k e^{i\omega k h}$$

is the spectral function of the operator $\boldsymbol{F} \cdot$.

The operator $\boldsymbol{F} \cdot$ is called a filter, inasmuch as its intent is to modify the set $\{u_n\}$ in a manner which is specified in the frequency domain by $\hat{F}(\omega)$. In engineering terms, $\hat{F}(\omega)$ would be called the "frequency response" of the filter.

If the filter is intended to modify only the amplitude, then $\hat{F}(\omega)$ must be real. This is obtained by choosing $\boldsymbol{F} \cdot$ to be symmetrical, i.e., $f_k = f_{-k}$. (Note: there is a large body of literature on digital filtering. See, e.g., Lanczos (1956) for an early, reasonably theoretical text and Hamming (1977) for a recent, more applied one.)

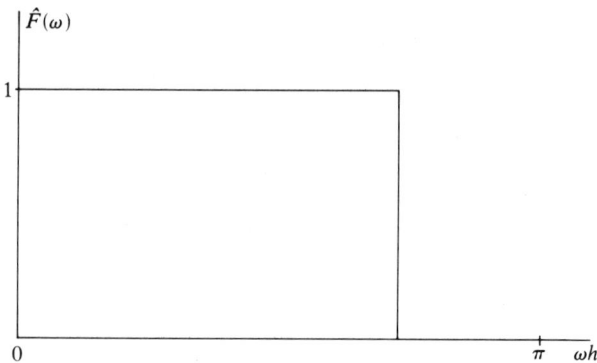

FIG. 5.3. *Ideal low pass filter.*

5.5. Low pass filters. In the desire to eliminate undesirable high frequency sinusoidal components of numerical solutions without affecting the low frequency ones, we are led to look for a numerical filter which has the ideal low pass amplitude response illustrated in Fig. 5.3, and has no effect on the phase.

While this cannot be realized simply,[1] such a filter may serve as an "ideal" which can be approximated by simple filters. Consider for example the family of symmetrical filters described by

$$(5.35) \qquad \boldsymbol{F} \cdot = a_0 + \frac{a_1}{2}(\boldsymbol{E} \cdot + \boldsymbol{E}^{-1} \cdot) + \frac{a_2}{2}(\boldsymbol{E}^2 \cdot + \boldsymbol{E}^{-2} \cdot) + \cdots .$$

That is, the operation of filtering is

$$(5.36) \qquad u_n^F = a_0 u_n + \frac{a_1}{2}(u_{n+1} + u_{n-1}) + \frac{a_2}{2}(u_{n+2} + u_{n-2}) + \cdots .$$

The amplitude response of this filter

$$(5.37) \qquad \hat{F}(\omega) = a_0 + a_1 \cos(\omega h) + a_2 \cos(2\omega h) + \cdots$$

is a real function for all ω.

If $\boldsymbol{F} \cdot$ (see (5.31)) were not symmetrical, then $\hat{F}(\omega)$ would contain an imaginary part, and the requirement that the phase of sinusoidal solutions not be affected would be violated.

5.6. Flat low pass filters. We may derive the undetermined coefficients in (5.35) by requiring the following conditions:

---

[1] See § 5.7.

(i) that zero frequency components be unaffected

(5.38) $$\hat{F}(0) = 1;$$

(ii) that components of wavelength $2h$ or $\omega = \pi/h$ be entirely eliminated

(5.39) $$\hat{F}\left(\frac{\pi}{h}\right) = 0;$$

(iii) that the remaining degrees of freedom be used to make $\hat{F}(\omega)$ as flat as possible in the neighborhood of $\omega = 0$,

(5.40) $$\frac{d}{d(\omega^2)}(\hat{F}(\omega))_0 = 0, \quad \frac{d^2}{d(\omega^2)^2}(\hat{F}(\omega))_0 = 0, \quad \cdots.$$

The equations which implement conditions (i) and (iii) are obtained by expanding the cosine terms of (5.37) in a MacLaurin series:

(5.41) $$\hat{F}(\omega) = a_0 + a_1\left(1 - \frac{(\omega h)^2}{2} + \frac{(\omega h)^4}{4!} - \cdots\right)$$
$$+ a_2\left(1 - \frac{(2\omega h)^2}{2} + \frac{(2\omega h)^4}{4!} - \cdots\right)$$
$$+ \cdots.$$

We thus obtain the equations
(i) $a_0 + a_1 + a_2 + \cdots = 1;$
(ii) $a_0 - a_1 + a_2 - \cdots = 0;$
(iii) $a_1 + 2^2 a_2 + 3^2 a_3 + \cdots = 0,$
$a_1 + 2^4 a_2 + 3^4 a_3 + \cdots = 0,$
$\vdots$

etc.

Table 5.1 lists numerical values of the a_k for filters in this class, (up to 11 points), and Fig. 5.4 gives the corresponding amplitude responses.

5.7. Fast Fourier transform filtering. An alternative form of filtering is provided by transforming the function to the Fourier domain where filtering becomes simple multiplication of the harmonic components. This may be achieved economically by use of the fast Fourier transform algorithm. Let

(5.42) $$\{u_n\}$$

Table 5.1
Coefficients a_k for flat low pass filters for 3-, 5-, 7-, 9- and 11-point formulae

Number of points used in filtering	Corresponding coefficients
3	$a_0 = 0.500000$ $a_1 = 0.500000$
5	$a_0 = 0.625000$ $a_1 = 0.500000$ $a_2 = -0.12500$
7	$a_0 = 0.687499$ $a_1 = 0.468750$ $a_2 = -0.187500$ $a_3 = 0.031250$
9	$a_0 = 0.726562$ $a_1 = 0.437500$ $a_2 = -0.218750$ $a_3 = 0.062500$ $a_4 = -0.007812$
11	$a_0 = 0.753906$ $a_1 = 0.410155$ $a_2 = -0.234375$ $a_3 = 0.087890$ $a_4 = -0.019531$ $a_5 = 0.001953$

be a *finite* set of values of a function $u(x)$ given in the $2N$ points on the x-axis,

(5.43) $$\{x_n = nh; n = 0, 1, 2, \cdots, 2N-1\}.$$

The discrete Fourier transform of $\{u_n\}$ is

(5.44) $$\bar{u}_k = h \sum_{n=0}^{2N-1} u_n e^{-ik\pi n/N}$$

and, conversely, (with $l = 2Nh$)

(5.45) $$u_n = \frac{1}{l} \sum_{k=-N}^{N} \bar{u}_k e^{ik\pi n/N}.$$

However, we may consider the shortened series

(5.46) $$u_n^F = \frac{1}{l} \sum_{k=-N+M}^{N-M} \bar{u}_k e^{ik\pi n/N}$$

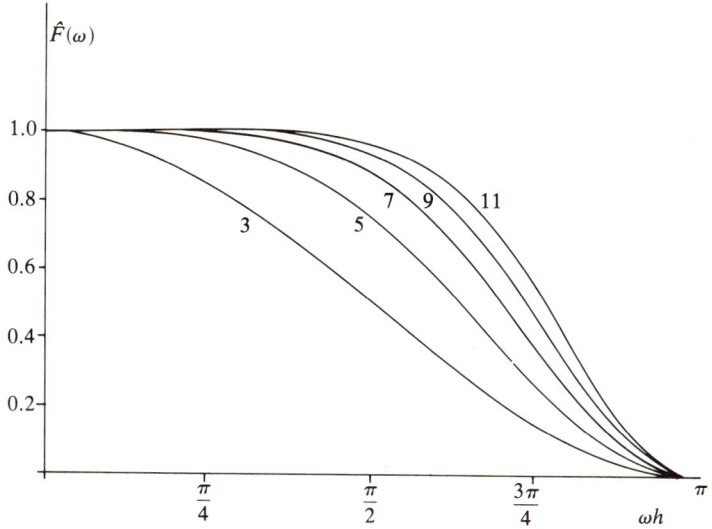

FIG. 5.4. *Amplitude response of flat low pass filters for 3-, 5-, 7-, 9- and 11-point formulae.*

(with $M < N$), from which the M highest harmonics in $\{\bar{u}_k\}$ have been simply deleted.

Equation (5.46) describes a filter which has, *in the discrete points $\omega_k = k\pi/Nh$*, the ideal frequency response shown in Fig. 5.5.

This filter can be implemented efficiently using the fast Fourier transform (FFT) algorithm as shown in Fig. 5.6. The transformations (5.34)–(5.35) as implemented in the FFT require on the order of $N \log(N)$ operations (as compared to the N^2 operations suggested by implementing the multiplications

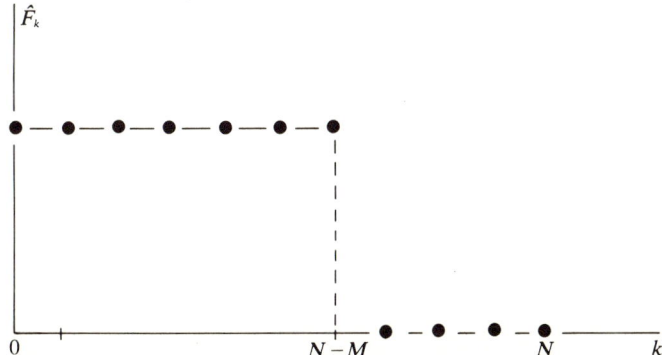

FIG. 5.5. *Discrete frequency response of the filter described by (5.46).*

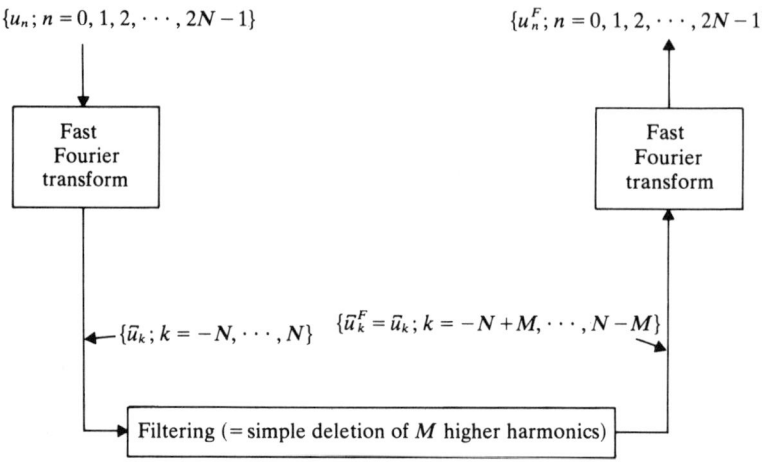

FIG. 5.6. *Filtering in the frequency domain using two fast Fourier transformations.*

and additions separately). Comments about the use of such filters, as applied to numerical simulations of geophysical fluid dynamics equations, may be found in Fornberg (1975), Kreiss and Oliger (1972) and Gottlieb and Orszag (1977). See also Vichnevetsky (1981c) for details about the fast Fourier transform and further references.

5.8. Effect of filtering on damping and numerical stability. Filtering may be applied to the numerical solution $\{u_n^j\}$ every r time steps (r need not be 1). The amplification factor of the calculation with filter is then, *on the average*

$$(5.47) \qquad z_F(\omega) = z(\omega)(\hat{F}(\omega))^{1/r}.$$

It is this expression which describes the damping per time step introduced by the filter. Likewise, the von Neumann condition of numerical stability is now

$$(5.48) \qquad |z_F(\omega)| = |z(\omega)||\hat{F}(\omega)|^{1/r} \leq 1 \quad \text{for all } |\omega| \leq \frac{\pi}{h}.$$

In the particular case of Fourier transform filtering, this becomes

$$(5.49) \qquad |z(\omega)| \leq 1 \quad \text{for } |\omega| \leq \frac{\pi}{h}\left(1 - \frac{M}{N}\right).$$

Chapter 6

Group Velocity

6.1. Introduction. The concept of group velocity was first introduced in mathematical physics to describe the propagation in a dispersive medium of "wave packets" or short wavelength oscillations modulating a relatively slowly varying envelope, as illustrated in Fig. 6.1.

A dispersive medium is a medium in which sinusoidal waves propagate at a (phase) velocity which is a function of the frequency ω or wavelength $2\pi/\omega$. Dispersive media in physics typically occur in liquids and solids which are transparent to light. The frequency dependent velocity is created here by the interaction between sinusoidal waves and the discrete structure, consisting of (on the average) equally spaced individual atoms.[1] The situation with numerical approximations is similar, in that sinusoidal wave solutions interacting

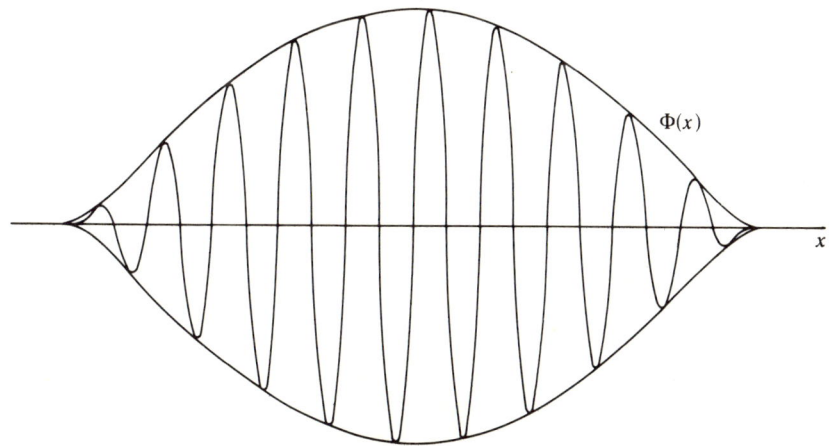

FIG. 6.1. *"Wave packet." Short wavelength oscillations modulating a slowly varying envelope.*

---

[1] See, e.g., Brillouin (1946).

with the regular discretization of the domain of the equation also have the effect of producing a frequency dependent velocity.

A function of the general shape of a "wave packet," as shown in Fig. 6.1, may be expressed as

(6.1) $$u(x, t) = \Phi(x, t) e^{i(\omega x + \mu(t))},$$

where $\Phi(x, t)$ is the envelope and $\mu(t)$ is the phase of the short wavelength modulation ($\mu = 0$ for $t = 0$).

We may assume that the initial envelope $\Phi(x, 0)$ is expressed in Fourier series form as

(6.2) $$\Phi(x, 0) = \sum_\gamma a_\gamma e^{i\gamma x},$$

where, for all γ, $\gamma \ll \omega$.

The initial function is thus

(6.3) $$u(x, 0) = \sum_\gamma a_\gamma e^{i(\gamma+\omega)x} = \sum_\gamma u_\gamma(x, 0),$$

where each component $u_\gamma(x, 0)$ is

(6.4) $$u_\gamma(x, 0) = a_\gamma e^{i(\gamma+\omega)x}.$$

Assume that, as before, $c^*(\omega)$ denotes the nonconstant phase velocity of sinusoidal components. (Its origin need not come from a numerical discretization.) We note that each $u_\gamma(x, 0)$ is sinusoidal in x. We may thus apply to each the appropriate phase velocity

(6.5) $$u_\gamma(x, t) = a_\gamma e^{i(\gamma+\omega)[x - c^*(\gamma+\omega)t]}.$$

It has been assumed that $\gamma \ll \omega$. We may thus rewrite the above, expanding the coefficient of t in a Taylor series with respect to γ and limiting the expansion to the linear term, as

(6.6) $$\begin{aligned} u_\gamma(x, t) &= a_\gamma e^{i[(\gamma+\omega)x - (\omega c^*(\omega) + \mathcal{V}(\omega)\gamma)t]} \\ &= a_\gamma e^{i\gamma(x - \mathcal{V}(\omega)t)} e^{i\omega(x - c^*(\omega)t)}. \end{aligned}$$

Here $\mathcal{V}(\omega)$, called the *group velocity*, is given by

(6.7) $$\mathcal{V}(\omega) = \frac{d}{d\omega}(\omega c^*(\omega)).$$

We may observe that (6.6) is *the expression of a function which has the phase velocity $c^*(\omega)$, and an envelope $a_\gamma \exp(i\gamma(x - \mathcal{V}t))$ which propagates without deformation at the velocity $\mathcal{V}(\omega)$*. Since the group velocity $\mathcal{V}(\omega)$ is independent of γ, any smooth initial envelope $\Phi(x, 0)$ which may be synthesized by a sum *of low frequency Fourier components of the form* (6.4) *will obey the same law of propagation*:

(6.8) $$\Phi(x, t) = \sum_\gamma u_\gamma(x, t) = \Phi(x - \mathcal{V}(\omega)t, 0).$$

It is important to note the fact that the group velocity (or velocity of the envelope) is a function of the modulating frequency ω. We may also note that when the phase velocity is constant ($c^*(\omega) = c =$ constant), then (6.7) yields simply $\mathcal{V} = c$.

The theory of group velocity is important in the present context mainly in explaining the propagation of short wavelength, spurious oscillations which appear near discontinuities in discrete approximations of hyperbolic equations.[2] These spurious oscillations typically have a $2h$ wavelength. Because of the nonconstancy of $c^*(\omega)$, the envelope of such spurious oscillations is observed to propagate at the group velocity $\mathcal{V}(\pi/h)$ rather than the phase velocity $c^*(\pi/h)$. As we shall see in the following examples, this group velocity is in general negative, whereas $c^*(\omega)$ is nonnegative for all ω in $[-\pi/h, \pi/h]$.

Spurious, short wavelength oscillations may appear in numerical solutions near boundaries. An example is given in § 7.7. They are also generated at the interface of mesh refinements (Vichnevetsky (1981b)).

6.2. Group velocity and energy propagation. The energy contained in the wave packet (6.1) is

(6.9) $$\|u\|_2^2 \equiv \int |u(x, t)|^2 \, dx = \int |\Phi(x, t)|^2 \, dx.$$

Since the envelope propagates without deformation at the group velocity $\mathcal{V}(\omega)$, the energy contained in the wave packet travels with the wave packet at the *group velocity* and not the phase velocity. (Therefore $\mathcal{V}$ is also called the energy velocity.)

This result is important. We shall see (in particular in Chapt. 7) that certain spurious numerical solutions of semi-discretizations are affected with a positive phase velocity but a *negative group velocity*. Such solutions thus transport energy of the numerical solution in the direction opposite to that of the flow.

6.3. Group velocity of the simple 3-point finite differences semi-discretization. For the central difference semi-discretization

(6.10) $$\frac{du_n}{dt} = -c \left(\frac{u_{n+1} - u_{n-1}}{2h} \right),$$

---

[2] See Vichnevetsky and Tomalesky (1971).

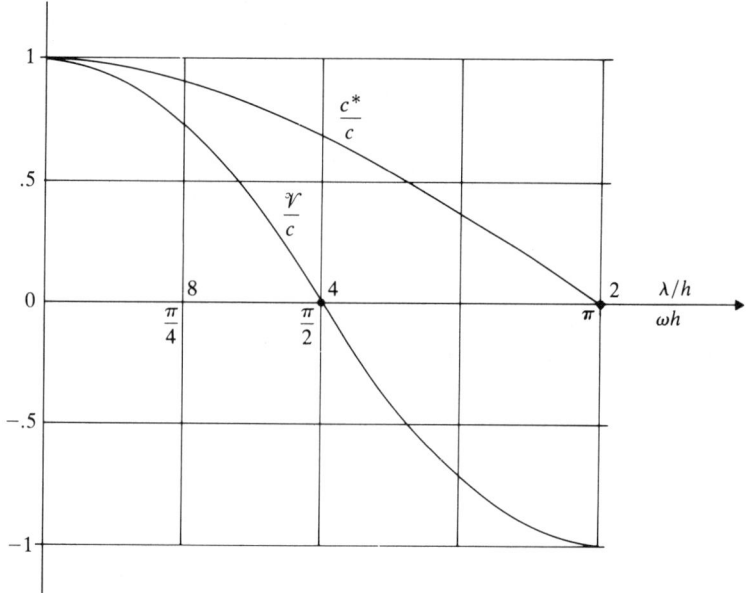

FIG. 6.2. *Phase and group velocity of the simple 3-point finite differences semi-discretization* (6.10).

the phase velocity is, from Table 2.1,

(6.10′) $$c^*(\omega) = c\left(\frac{\sin(\omega h)}{\omega h}\right).$$

Thus, by (6.7) the group velocity is

(6.11) $$\mathcal{V}(\omega) = c \cos(\omega h).$$

This function is shown in Fig. 6.2. Note in particular that the group velocity for $\omega = \pi/h$ (corresponding to a wavelength $\lambda = 2\pi/\omega = 2h$) is

(6.12) $$\mathcal{V}\left(\frac{\pi}{h}\right) = -c.$$

6.4. Group velocity of other 3-point semi-discretizations. We derived in (2.34) the expression

(6.13) $$c^*(\omega) = c\left(\frac{1}{(1-\beta) + \beta \cos(\omega h)}\right)\left(\frac{\sin(\omega h)}{\omega h}\right)$$

for the phase velocity of the family of 3-point finite element/weighted residual semi-discretizations (see (2.33))

(6.14) $$\left(\frac{\beta}{2}\frac{du_{n-1}}{dt} + (1-\beta)\frac{du_n}{dt} + \frac{\beta}{2}\frac{du_{n+1}}{dt}\right) = -c\left(\frac{u_{n+1} - u_{n-1}}{2h}\right).$$

The corresponding group velocity is thus found by (6.7) to be

(6.15) $$\mathcal{V}(\omega) = c\frac{\beta + (1-\beta)\cos(\omega h)}{((1-\beta) + \beta\cos(\omega h))^2}.$$

This function is illustrated in Fig. 6.3 for several values of β. We find in particular that the group velocity of the $2h$ wavelength solutions is

(6.16) $$\mathcal{V}\left(\frac{\pi}{h}\right) = c\left(\frac{-1}{1-2\beta}\right),$$

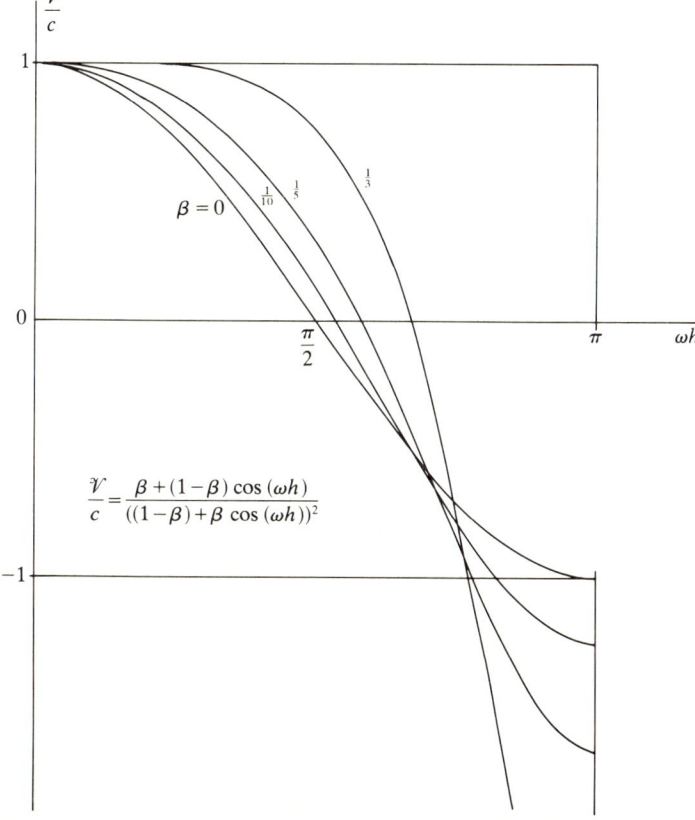

FIG. 6.3. *Group velocity of the family of 3-point finite element/weighted residual semi-discretizations* (6.14) *for various values of β.*

which is seen to be negative for all values of $\beta < \frac{1}{2}$. For $\beta = 0$ (which is the central difference case (6.10)) we find as before that $\mathcal{V}(\pi/h) = -c$.

Note that the mean value of $\mathcal{V}(\omega)$ is zero (for all β)

$$(6.17) \qquad \int_0^{\pi/h} \mathcal{V}(\omega)\, d\omega = \int_0^{\pi/h} \frac{d}{d\omega}(\omega c^*(\omega))\, d\omega = [\omega c^*(\omega)]_0^{\pi/h} = 0.$$

6.5. Wave analysis. Some of the preceding results may be derived from a completely different approach. Consider as before the central differences semi-discretization (6.10) and relabel the variables u_n as v_n for n odd. Hence, the equation takes the form

$$(6.18) \qquad \frac{du_n}{dt} = -c\left(\frac{v_{n+1} - v_{n-1}}{2h}\right) \quad \text{for } n \text{ even,}$$

$$(6.19) \qquad \frac{dv_n}{dt} = -c\left(\frac{u_{n+1} - u_{n-1}}{2h}\right), \quad \text{for } n \text{ odd.}$$

These equations are seen to be consistent difference approximations of

$$(6.20\text{a}) \qquad \frac{\partial u}{\partial t} = -c\frac{\partial v}{\partial x},$$

$$(6.20\text{b}) \qquad \frac{\partial v}{\partial t} = -c\frac{\partial u}{\partial x}.$$

By eliminating v, this system may be transformed into

$$(6.21) \qquad \frac{\partial^2 u}{\partial t^2} - c^2 \frac{\partial^2 u}{\partial x^2} = 0,$$

which shows that *the central difference semi-discretization* (6.10) *is a consistent approximation of the wave equation* (6.21) rather than of the advection equation $U_t + cU_x = 0$. To the wave equation are associated a *forward wave* solution and a *backward wave* solution. While the forward solution is that which approximates solutions of the advection equation, we will show that the backward solution consists of waves of wavelength $2h$, traveling at the velocity $-c$ which appeared as a group velocity in the preceding analysis. To show this, we add and subtract the two equations in (6.20) to obtain the characteristic form

$$(6.22\text{a}) \qquad \frac{\partial}{\partial t}\left(\frac{u+v}{2}\right) + c\frac{\partial}{\partial x}\left(\frac{u+v}{2}\right) = 0,$$

$$(6.22\text{b}) \qquad \frac{\partial}{\partial t}\left(\frac{u-v}{2}\right) - c\frac{\partial}{\partial x}\left(\frac{u-v}{2}\right) = 0,$$

or with

$$w_1(x, t) = \frac{u+v}{2}, \quad w_2(x, t) = \frac{u-v}{2}$$

also

(6.23a) $$\frac{\partial w_1}{\partial t} + c \frac{\partial w_1}{\partial x} = 0,$$

(6.23b) $$\frac{\partial w_2}{\partial t} - c \frac{\partial w_2}{\partial x} = 0.$$

The characteristic velocity of (6.23a) is $+c$, while that of (6.23b) is $-c$.

The implication of these results is as follows. The forward wave $w_1(x, t)$ represents locally the mean of two successive values, and constitutes therefore the smooth part of the numerical solution. According to (6.23a), it is seen to propagate at the (correct) velocity $+c$.

By constrast, the function $w_2(x, t)$ is seen, by (6.23b), to propagate *backward* at the velocity $-c$. Since (see Fig. 6.4)

$$w_2 = \frac{u-v}{2} = u - \left(\frac{u+v}{2}\right),$$

w_2 is seen to represent the oscillatory part of the solution (of wavelength $2h$). It represents a spurious solution or *error wave* of the difference approximation, and is precisely the kind of "wave packet," traveling at the group velocity $\mathscr{V}(\pi/h) = -c$ which was found before.

Figure 6.5 shows the time evolution of $(u+v)$ and $(u-v)$ waves for several initial functions of varying degree of smoothness. These were obtained by numerical integration of the semi-discretization (6.10). Time marching was implemented with the leapfrog method, using the value $c\Delta t/h = 0.2$ for the

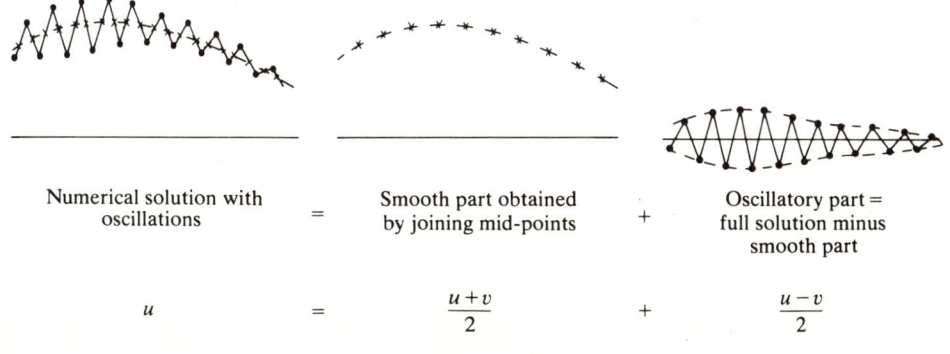

FIG. 6.4.

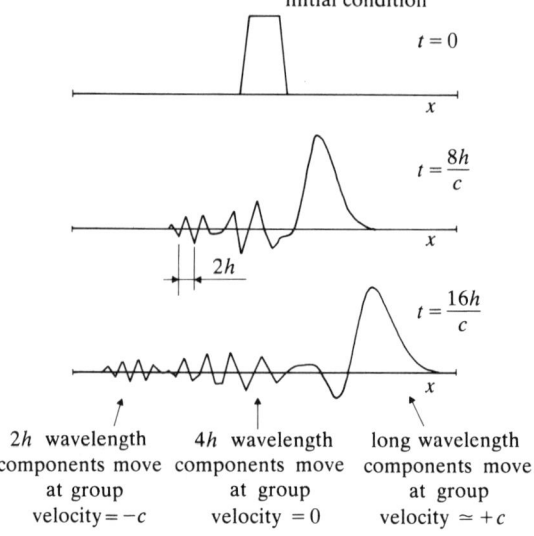

FIG. 6.5a. *Numerical experiment: mostly $(u+v)$ wave moving forward at velocity $+c$.*

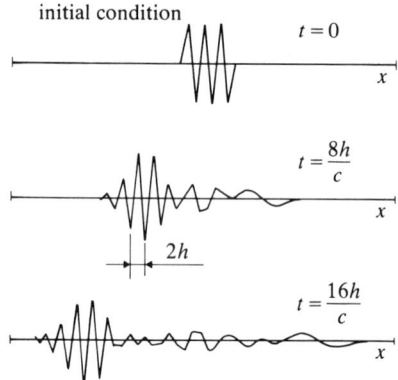

FIG. 6.5b. *Numerical experiment: mostly $(u-v)$ wave moving backward at velocity $-c$.*

Courant number. With this value of $c\Delta t/h$, the error is almost entirely due to the semi-discretization in space, and the numerical solutions shown in Fig. 6.5 are as predicted by the group velocity theory described here.

6.6. Wave analysis of other semi-discretizations. These results are easily generalized to other cases: Consider the general 3-point semi-discretization

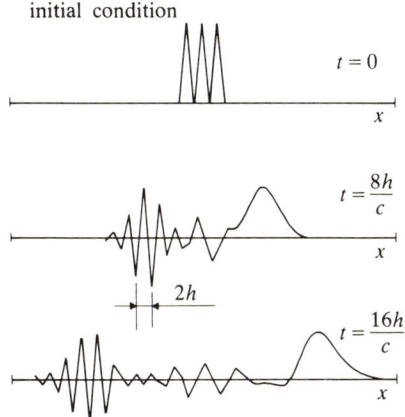

FIG. 6.5c. *Numerical experiment: superposition of a $(u+v)$ wave and a $(u-v)$ wave moving in opposite directions at velocity $+c$ and $-c$ respectively.*

(2.33). Proceeding as before, we relabel every other variable as v_n, and obtain

(6.24)
$$\frac{\beta}{2}\left(\frac{dv_{n+1}}{dt}+\frac{dv_{n-1}}{dt}\right)+(1-\beta)\frac{du_n}{dt}=-c\left(\frac{v_{n+1}-v_{n-1}}{2h}\right),$$
$$\frac{\beta}{2}\left(\frac{du_{n+1}}{dt}+\frac{du_{n-1}}{dt}\right)+(1-\beta)\frac{dv_n}{dt}=-c\left(\frac{u_{n+1}-u_{n-1}}{2h}\right),$$

which is a consistent approximation of

(6.25)
$$\beta\frac{\partial v}{\partial t}+(1-\beta)\frac{\partial u}{\partial t}=-c\frac{\partial v}{\partial x},$$
$$\beta\frac{\partial u}{\partial t}+(1-\beta)\frac{\partial v}{\partial t}=-c\frac{\partial u}{\partial x}.$$

This system is easily transformed into its characteristic form

(6.26)
$$\frac{\partial}{\partial t}\left(\frac{u+v}{2}\right)=-c\frac{\partial}{\partial x}\left(\frac{u+v}{2}\right),$$
$$\frac{\partial}{\partial t}\left(\frac{u-v}{2}\right)=c\left(\frac{1}{1-2\beta}\right)\frac{\partial}{\partial x}\left(\frac{u-v}{2}\right),$$

showing that the "smooth" $w_1=(u+v)/2$ component of the solution travels at the expected velocity $+c$, while the "oscillatory" part $w_2=(u-v)/2$ travels at the velocity

(6.27)
$$-c\left(\frac{1}{1-2\beta}\right)$$

which is equal to the group velocity $\mathscr{V}(\pi/h)$ found in (6.16).

Chapter 7

Time-Fourier Transforms

7.1. Introduction. The main tool used so far has been Fourier transforms in the x-direction (or x-Fourier transforms). We have been able to analyze in considerable detail the propagation properties of regular numerical approximations of hyperbolic equations in domains assumed to be infinite. But x-Fourier transforms become inadequate to analyze what happens near discontinuities in the computational domain, such as boundaries (which must always be dealt with in practice) and mesh refinement (which is often used in fluid dynamics calculations requiring a greater accuracy in specified subdomains). We will show in this chapter that time-Fourier transforms (the independent variable in the transformed domain is the time-frequency Ω having dimensions t^{-1}) reveal new properties of numerical solutions, and provide the needed tool for the analysis of their behavior near discontinuities of the computational domain. As before, we examine the semi-discretization

(7.1) $$\frac{du_n}{dt} = -c\left(\frac{u_{n+1} - u_{n-1}}{2h}\right)$$

of the simple equation

(7.2) $$\frac{\partial U}{\partial t} + c\frac{\partial U}{\partial x} = 0$$

taken as a model of hyperbolic equations.

Let $\{\hat{u}_n(\Omega)\}$ denote the set of time-Fourier transforms (or t-Fourier transforms) of the semi-discrete numerical solution $\{u_n\}$, defined as

(7.3) $$\hat{u}_n(\Omega) = \int_{-\infty}^{\infty} u_n(t)\, e^{-i\Omega t}\, dt.$$

The converse relation is

(7.4) $$u_n(t) = \int_{-\infty}^{\infty} \hat{u}_n(\Omega)\, e^{i\Omega t}\frac{d\Omega}{2\pi}.$$

We take the time-Fourier transform of (7.1) and obtain

(7.5) $$i\Omega \hat{u}_n = -c\left(\frac{\hat{u}_{n+1} - \hat{u}_{n-1}}{2h}\right)$$

or

(7.6) $$\hat{u}_{n+1} + 2i\left(\frac{\Omega h}{c}\right)\hat{u}_n - \hat{u}_{n-1} = 0.$$

Solving this recurrence equation for $\{\hat{u}_n\}$ may be achieved by seeking "normal" or "fundamental" solutions, i.e., solutions for which the ratio

(7.7) $$\frac{\hat{u}_{n+1}}{\hat{u}_n} \equiv \hat{E}(\Omega)$$

is independent of n. We note that $\hat{E}$ is an image in the time-Fourier domain of the space shift operator $\mathbf{E}$ · defined in Chapter 1.

To find solutions that obey (7.7) we insert this expression in (7.6) and obtain

(7.8) $$\left(\hat{E} + 2i\left(\frac{\Omega h}{c}\right) - \hat{E}^{-1}\right)\hat{u}_n = 0,$$

indicating that $\hat{E}(\Omega)$ must satisfy the characteristic equation

(7.9) $$\hat{E}^2 + 2i\left(\frac{\Omega h}{c}\right)\hat{E} - 1 = 0.$$

This equation has the two *characteristic roots* (see Fig. 7.1):

(7.10) $$\hat{E}_1(\Omega) = -i\left(\frac{\Omega h}{c}\right) + \sqrt{1 - \left(\frac{\Omega h}{c}\right)^2},$$

(7.11) $$\hat{E}_2(\Omega) = -i\left(\frac{\Omega h}{c}\right) - \sqrt{1 - \left(\frac{\Omega h}{c}\right)^2}.$$

Thus two types of fundamental solutions may exist satisfying

(7.12) $$\frac{\hat{u}_{n+1}}{\hat{u}_n} = \hat{E}_1 \quad \text{and} \quad \frac{\hat{u}_{n+1}}{\hat{u}_n} = \hat{E}_2,$$

respectively. This may be stated as

PROPERTY 1. *Numerical solutions of* (7.1) *may be expressed as the sum*

(7.13) $$\{u_n(t)\} = \{p_n(t)\} + \{q_n(t)\}$$

of two fundamental types of solution. Their respective Fourier transforms

(7.14) $$\hat{p}_n(\Omega) = \hat{p}_0(\Omega)[\hat{E}_1(\Omega)]^n$$

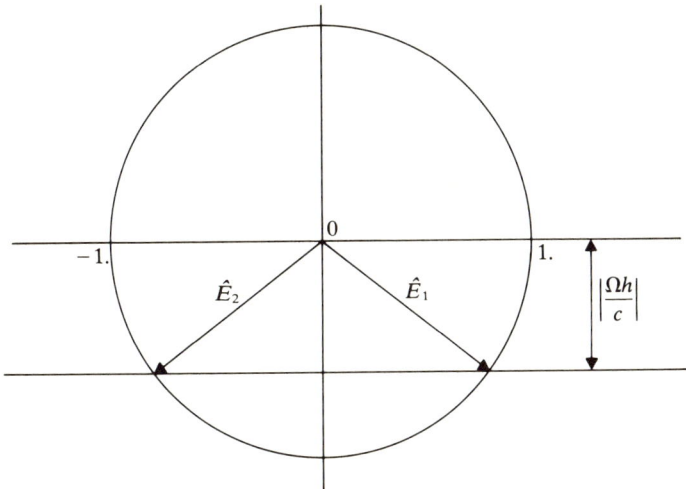

FIG. 7.1. *Roots of the characteristic equation* (7.9) *when* $\Omega h/c \leq 1$.

and

(7.15) $$\hat{q}_n(\Omega) = \hat{q}_0(\Omega)[\hat{E}_2(\Omega)]^n$$

describe the different propagation properties that apply to the two types.

That *two* types of solutions may exist is a consequence of the fact that (as we have seen in § 6.5), the semi-discretization (7.1) is a consistent approximation of the *second-order* wave equation

(7.16) $$\frac{\partial^2 U}{\partial t^2} - c^2 \frac{\partial^2 U}{\partial x^2} = 0$$

rather than the first order equation (7.2).

7.2. Numerical phase velocity and wavelength. For comparison with (7.13), we express exact solutions $U(x, t)$ of the advection equation also in time-Fourier integral form

(7.17) $$\hat{U}(x, \Omega) = \int_{-\infty}^{\infty} U(x, t) e^{-i\Omega t} dt.$$

If $U(x, t)$ is to be a solution of (7.2), then $\hat{U}(x, \Omega)$ must satisfy

(7.18) $$i\Omega \hat{U} + c \frac{\partial \hat{U}}{\partial x} = 0$$

or, by analytic integration,

(7.19) $$\hat{U}(x, \Omega) = \hat{U}(0, \Omega) e^{-i\Omega x/c}.$$

For each Ω the x-dependence is harmonic (sinusoidal),
- with a conservative amplitude

(7.20) $$|\hat{U}(x, \Omega)| = |\hat{U}(0, \Omega)|,$$

- with a spatial frequency ω given by

(7.21) $$\omega = \frac{\Omega}{c},$$

- and with a wavelength λ given by

(7.22) $$\lambda = \frac{2\pi}{\omega} = \frac{2\pi c}{\Omega}.$$

The numerical equivalent of (7.19) is expressed by (7.14) and (7.15). When Ω is below the *cut-off* frequency

(7.23) $$\Omega_c = \frac{c}{h}$$

then we may verify that

(7.24) $$|\hat{E}_1(\Omega)| = |\hat{E}_2(\Omega)| = 1.$$

Then,

(7.25) $$|\hat{p}_n(\Omega)| = |\hat{p}_0(\Omega)[\hat{E}_1(\Omega)]^n| = |\hat{p}_0(\Omega)|$$

and

(7.26) $$|\hat{q}_n(\Omega)| = |\hat{q}_0(\Omega)[\hat{E}_2(\Omega)]^n| = |\hat{q}_0(\Omega)|.$$

In this band of time frequencies, the variation of Fourier components is thus constant-amplitude sinusoidal in x as well as in t. To find the corresponding numerical phase velocities, we write, in a form analogous to (7.19),

(7.27) $$\hat{p}_n = \hat{p}_0 \, e^{n\angle \hat{E}_1} \equiv \hat{p}_0 \, e^{-i\Omega x_n/c_1^*(\Omega)},$$

(7.28) $$\hat{q}_n = \hat{q}_0 \, e^{n\angle \hat{E}_2} \equiv \hat{q}_0 \, e^{-i\Omega x/c_2^*(\Omega)},$$

where

(7.29) $$\angle \hat{E}_1(\Omega) = -\arcsin\left(\frac{\Omega h}{c}\right)$$

is the phase of $\hat{E}_1$, and where

(7.30) $$c_1^*(\Omega) = \frac{\Omega h}{\arcsin(\Omega h/c)}$$

is the phase velocity at which solutions of $\{p_n\}$ type propagate. The wavelength is given by a relation similar to (7.22)

$$(7.31) \qquad \lambda_1 = \frac{2\pi c_1^*}{\Omega} = \frac{2\pi h}{\arcsin(\Omega h/c)}.$$

Fundamental numerical solutions of the $\{q_n\}$ type have a smaller phase velocity,

$$(7.32) \qquad c_2^*(\Omega) = -\frac{\Omega h}{\angle \hat{E}_2} = \frac{\Omega h}{\pi - \arcsin(\Omega h/c)},$$

and a correspondingly shorter wavelength,

$$(7.33) \qquad \lambda_2 = \frac{2\pi c_2^*}{\Omega} = \frac{2\pi h}{\pi - \arcsin(\Omega h/c)}.$$

These are illustrated in Figs. 7.2 and 7.3.

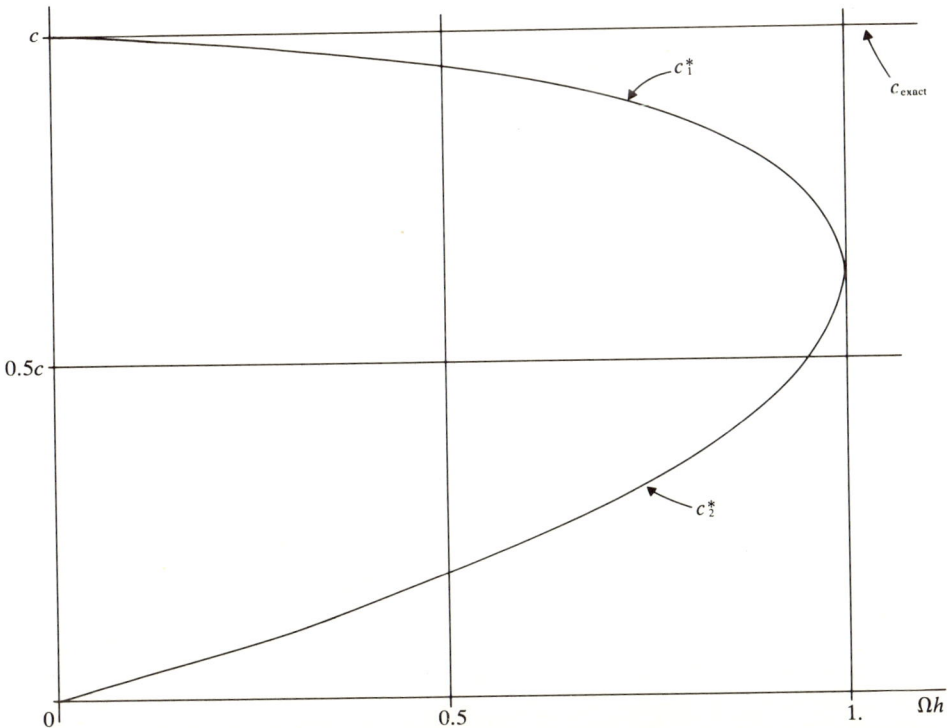

FIG. 7.2. *Phase velocity of the semi-discretization* (7.1).

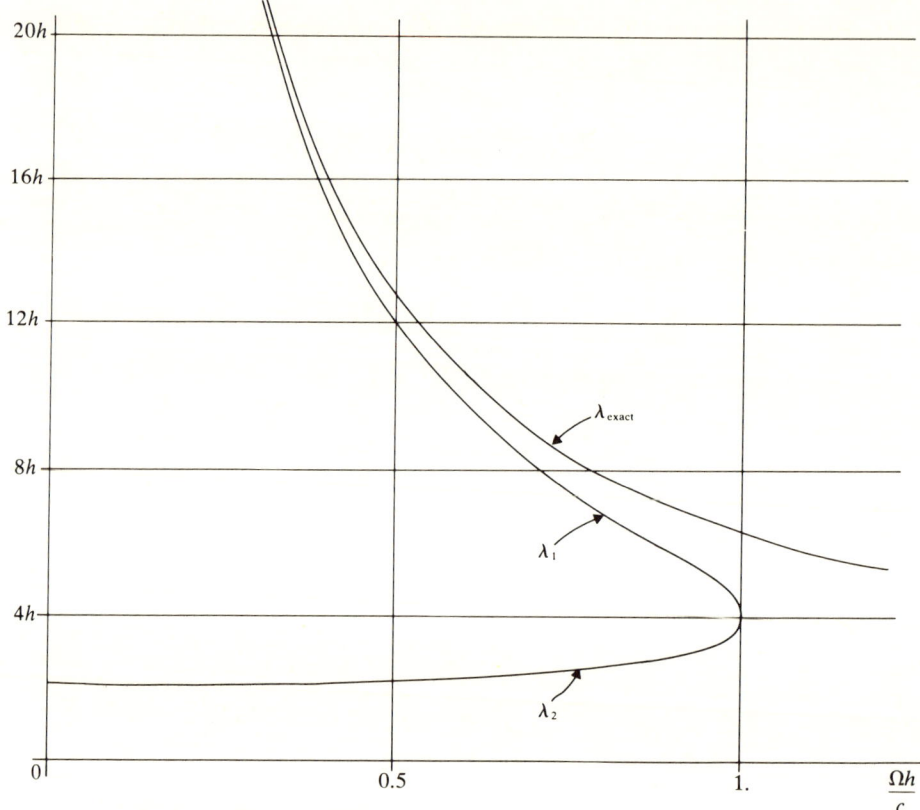

FIG. 7.3. *Exact and numerical wavelengths corresponding to a time frequency* Ω.

7.3. Relation to x-Fourier transforms. It is interesting at this point to return to x-Fourier analysis and to examine how solutions of $\{p_n\}$ and $\{q_n\}$ type are characterized there. From relation (2.44) repeated here,

$$(7.34) \qquad \omega = \frac{\Omega}{c^*(\Omega)},$$

we obtain by using (7.30) and (7.32)

$$(7.35a) \qquad \omega_1 = \frac{\Omega}{c_1^*(\Omega)} = \frac{1}{h} \arcsin\left(\frac{\Omega h}{c}\right),$$

which never exceeds $\pi/2h$, and

$$(7.35b) \qquad \omega_2 = \frac{\Omega}{c_2^*(\Omega)} = \frac{1}{h}\left(\pi - \arcsin\left(\frac{\Omega h}{c}\right)\right),$$

which is always greater than or equal to $\pi/2h$. The converse relation

$$\Omega = \frac{c}{h}\sin(\omega h)$$

holds for both types of fundamental solutions.

The domains in which ω and Ω take their values are $[-\pi/h, \pi/h]$ and $[-c/h, c/h]$ respectively. The correspondence is not one-to-one but two-to-two, viz., from (7.35):

- *for solutions of $\{p_n\}$ type,*

(7.36a) $$|\omega_1| \in \left[0, \frac{\pi}{2h}\right] \to |\Omega| \in \left[0, \frac{c}{h}\right];$$

- *for solutions of $\{q_n\}$ type,*

(7.36b) $$|\omega_2| \in \left[\frac{\pi}{2h}, \frac{\pi}{h}\right] \to |\Omega| \in \left[\frac{c}{h}, 0\right].$$

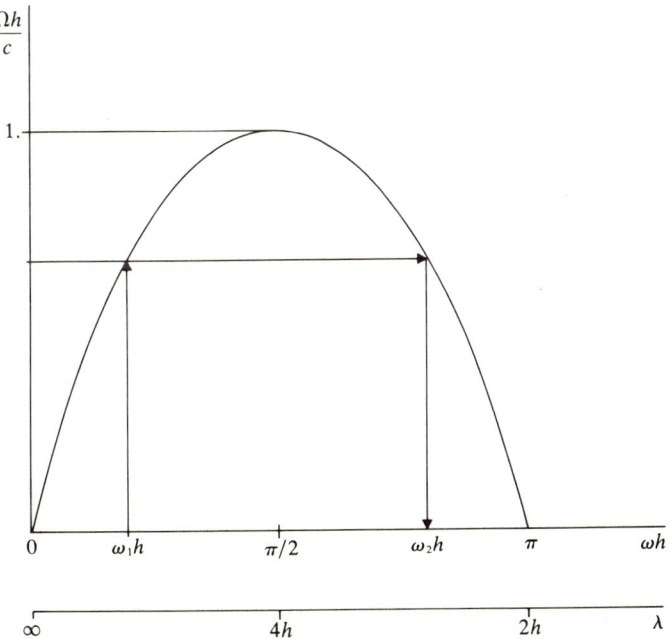

FIG. 7.4. *To each time frequency $\Omega < c/h$ correspond two space frequencies ω_1 and ω_2, characterizing fundamental solutions of $\{p_n\}$ and $\{q_n\}$ type, respectively. In reflection phenomena (§ 7.7), a numerical solution of frequency ω_1 generates a reflected solution of the corresponding frequency ω_2.*

We may thus express:

PROPERTY 2. *The x-Fourier transforms of fundamental solutions of $\{p_n\}$ and $\{q_n\}$ type have nonoverlapping support in ω:*

(7.37) $$0 \leq \omega_1 \leq \frac{\pi}{2h} \leq \omega_2 \leq \pi/h.$$

The same separation holds for wavelengths:

PROPERTY 3. *Numerical solutions of $\{p_n\}$ type have Fourier components with wavelengths λ_1 in $[4h, \infty)$, while numerical solutions of $\{q_n\}$ type have Fourier components with wavelengths λ_2 in $[2h, 4h]$.*

Phase velocities in terms of ω may be derived from (7.35a):

$$\omega = \frac{1}{h} \arcsin\left(\frac{\Omega h}{c}\right) = \frac{1}{h} \arcsin\left(\frac{\omega c^* h}{c}\right)$$

which reproduces the known result

(7.38) $$c^*(\omega) = c \frac{\sin(\omega h)}{\omega h}$$

that holds for both types of fundamental solutions.

7.4. Cut-off frequency. The situation is different when $|\Omega h/c| > 1$. Then the roots of the characteristic equation become (Fig. 7.5)

(7.39a) $$\hat{E}_1(\Omega) = -i\left(\frac{\Omega h}{c} - \sqrt{\left(\frac{\Omega h}{c}\right)^2 - 1}\right),$$

(7.39b) $$\hat{E}_2(\Omega) = -i\left(\frac{\Omega h}{c} + \sqrt{\left(\frac{\Omega h}{c}\right)^2 - 1}\right).$$

Both are pure imaginary, and neither is equal to one in absolute value. We then have

(7.40) $$|\hat{E}_1(\Omega)| < 1, \qquad |\hat{E}_2(\Omega)| > 1$$

with

$$|\hat{E}_1| \cdot |\hat{E}_2| = 1.$$

Their phase is

(7.41) $$\angle \hat{E}_1 = \angle \hat{E}_2 = -\frac{\pi}{2}.$$

The corresponding fundamental solutions of $\{p_n\}$ and $\{q_n\}$ type have envelopes that decay exponentially with x for the former (see Fig. 7.6) and with $-x$ for the latter. Such solutions cannot be generated by initial conditions on $x \in (-\infty, \infty)$, but may exist near boundaries of the computational domain.

An interesting property of those solutions is that their wavelength is equal to $4h$, irrespective of their frequency $\Omega > c/h$.

One situation in which such solutions may occur is in the approximation of (7.2) over the semi-infinite domain $x \geq 0$. The response to the simple sinusoidal boundary condition

(7.42) $$u_0(t) = U(0, t) = e^{i\Omega t}$$

is the numerical solution in $x > 0$:

(7.43) $$u_n(t) = e^{i\Omega t}[\hat{E}_1(\Omega)]^n = e^{i\Omega(t - x_n/c_1^*(\Omega))}|\hat{E}_1(\Omega)|^n.$$

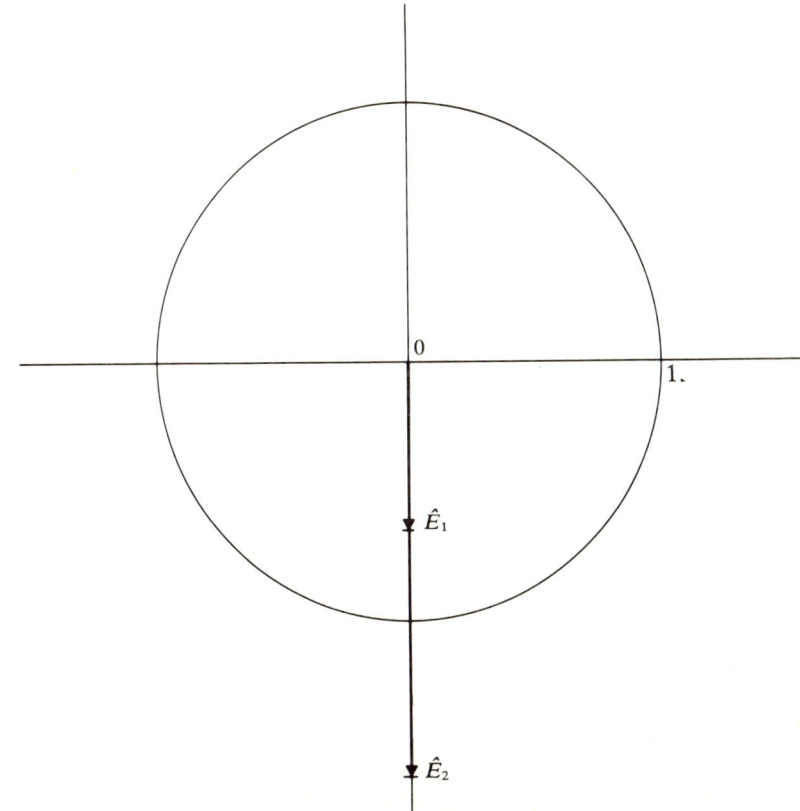

FIG. 7.5. *Roots of the characteristic equation (7.9) when $|\Omega h/c| > 1$.*

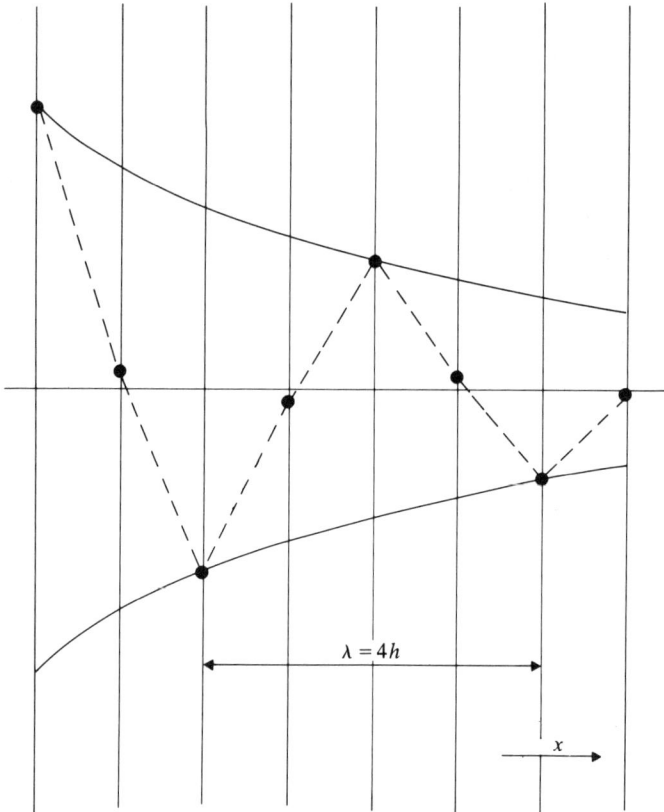

FIG. 7.6. *Numerical sinusoidal response when* $\Omega h/c > 1$. *There is amplitude decay with x as* $|\Omega h/c - \sqrt{(\Omega h/c)^2 - 1}|^n$.

When Ω exceeds the cut-off frequency $\Omega_c = c/h$, then (7.43) describes solutions that decay exponentially with x as

(7.44) $$|u_n| = |u_0| \cdot \left| \frac{\Omega h}{c} - \sqrt{\left(\frac{\Omega h}{c}\right)^2 - 1} \right|^n.$$

Such a solution is illustrated in Fig. 7.5. The reader may find in Roache (1972, pp. 56–58) numerical illustrations of this special case.

We shall see below, in § 7.6, that the cut-off frequency is also that for which the group velocity vanishes. Thus numerical solutions for $\Omega \geq \Omega c$ cannot carry energy away from the boundary $x = 0$ into the computational domain $x > 0$, whence the resulting amplitude decay.[1]

---

[1] See also remarks about this phenomenon in Browning, Kreiss and Oliger (1973).

7.5. Energy. We know from the analysis of previous chapters that solutions of the advection equation on $x \in (-\infty, \infty)$ are energy conservative

$$\frac{d}{dt}\|U(x,t)\|_2^2 \equiv \frac{d}{dt}\int_{-\infty}^{\infty} [U(x,t)]^2\, dx = 0, \tag{7.45}$$

and that this property is preserved by the semi-discretization (7.1), i.e.,

$$\frac{d}{dt}\|u_n\|_2^2 \equiv \frac{d}{dt}\left(h \sum u_n^2\right) = 0. \tag{7.46}$$

The energy of U may be expressed as a function of its Fourier transform by Parseval's relation as

$$\|U(x,t)\|_2^2 = \int_{-\infty}^{\infty} |\hat{U}(\omega,t)|^2 \frac{d\omega}{2\pi}. \tag{7.47}$$

Conservation of energy results here from the fact that

$$\frac{\partial}{\partial t}|\hat{U}(\omega,t)| = 0$$

for solutions of $U_t + cU_x = 0$. The energy of the discrete set $\{u_n\}$ is expressed by (1.62), the appropriate form of Parseval's relation, as

$$\|u_n\|_2^2 = \int_{-\pi/h}^{\pi/h} |\bar{u}(\omega,t)|^2 \frac{d\omega}{2\pi} = \int_0^{\pi/h} |\bar{u}(\omega,t)|^2 \frac{d\omega}{\pi}. \tag{7.48}$$

We may decompose the discrete Fourier transform $\bar{u}$ into its two components:

$$\bar{u}(\omega, t) = \bar{p}(\omega, t) + \bar{q}(\omega, t). \tag{7.49}$$

Then, from

$$\|p_n\|_2^2 = \int_0^{\pi/h} |\bar{p}|^2 \frac{d\omega}{\pi} = \int_0^{\pi/2h} |\bar{p}|^2 \frac{d\omega}{\pi} = \int_0^{\pi/2h} |\bar{u}|^2 \frac{d\omega}{\pi} \tag{7.50}$$

and

$$\|q_n\|_2^2 = \int_0^{\pi/h} |\bar{q}|^2 \frac{d\omega}{\pi} = \int_{\pi/2h}^{\pi/h} |\bar{q}|^2 \frac{d\omega}{\pi} = \int_{\pi/2h}^{\pi/h} |\bar{u}|^2 \frac{d\omega}{\pi}, \tag{7.51}$$

results the simple and interesting energy separation property:

PROPERTY 4. *The energy of numerical solutions $\{u_n\} = \{p_n\} + \{q_n\}$ satisfies the separation principle*

$$\|u_n\|_2^2 = \|p_n\|_2^2 + \|q_n\|_2^2. \tag{7.52}$$

7.6. Group velocity of the two fundamental types of solutions. The expression of the group velocity from $c^*(\Omega)$ (instead of $c^*(\omega)$) is easily derived.

From (6.7) and (7.34) we find

$$\mathcal{V} = \frac{d}{d\omega}(\omega c^*(\omega)) = \frac{d\Omega}{d(\Omega/c^*(\Omega))}.$$

Thus,

(7.53)
$$\frac{1}{\mathcal{V}} = \frac{d}{d\Omega}\left(\frac{\Omega}{c^*(\Omega)}\right).$$

If we apply this to the propagation of numerical solutions of $\{p_n\}$ type for which c^* is given by (7.30), then we find that

$$\frac{1}{\mathcal{V}_1} = \frac{d}{d\Omega}\left(\frac{\Omega}{c_1^*(\Omega)}\right) = \frac{1}{c\sqrt{1-(\Omega h/c)^2}}$$

or

(7.54)
$$\mathcal{V}_1 = c\sqrt{1-\left(\frac{\Omega h}{c}\right)^2},$$

which is positive for all $|\Omega h/c| < 1$. This may be stated as:

PROPERTY 5. *Numerical solutions of $\{p_n\}$ type (i.e., of wavelength $\lambda_1 > 4h$ or frequency $\omega_1 < \pi/2h$) have a positive group velocity. For smooth numerical solutions, $(\omega h \to 0$ or $\lambda/h \to \infty)$ this group velocity approaches $+c$.*

Smooth numerical solutions converge to the exact solution of (7.2) with both group and phase velocity approaching the exact value $+c$. Deriving now the group velocity of solutions of $\{q_n\}$ type gives

$$\frac{1}{\mathcal{V}_2} = \frac{d}{d\Omega}\left(\frac{\Omega}{c_2^*(\Omega)}\right) = \frac{-1}{c\sqrt{1-(\Omega h/c)^2}}$$

or

(7.55)
$$\mathcal{V}_2 = -c\sqrt{1-\left(\frac{\Omega h}{c}\right)^2}.$$

We observe that *the group velocity $\mathcal{V}_2$ of fundamental solutions of $\{q_n\}$ type is always negative.* The envelope of solutions characterized by $\{q_n\}$, and the energy carried by such solutions, thus travels backward in the computational domain.

PROPERTY 6. *Numerical solutions of $\{q_n\}$ type (i.e., of wavelength in the interval $2h \geq \lambda > 4h$) have a negative group velocity. For the highest frequency $\omega = \pi/h$ or $\lambda = 2h$) this group velocity equals $-c$.*

Numerical solutions of $\{q_n\}$ type of wavelength near $2h$ are precisely those described by $w_2(x, t)$ in the wave analysis of § 6.5, and their group velocity $\mathcal{V}_2(0) = -c$ in the frequency analysis is the analogue of the characteristic velocity $-c$ in the wave analysis. Such numerical solutions are consistent

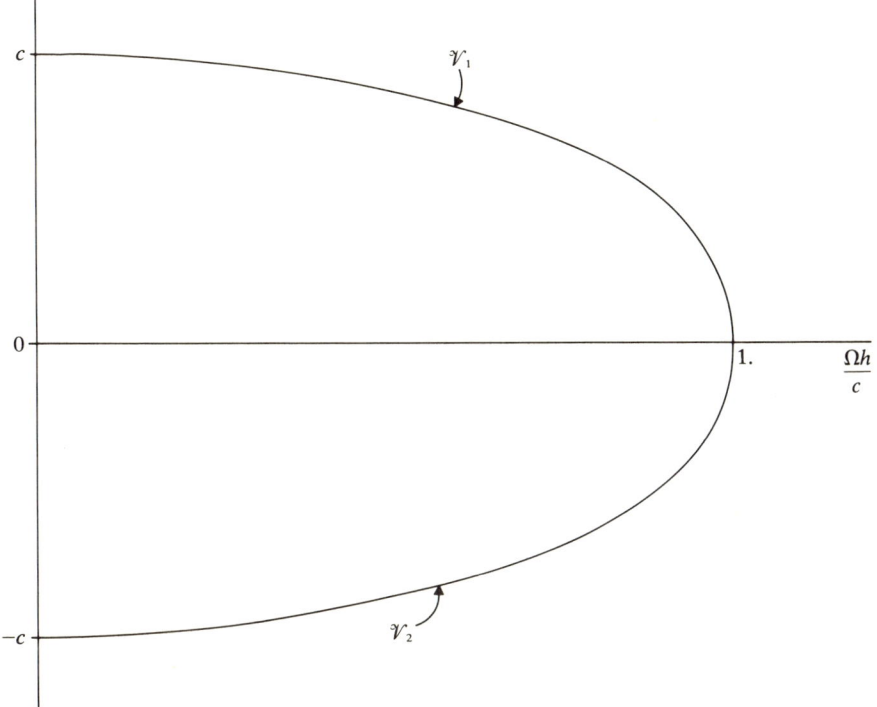

FIG. 7.7. *Group velocity of the finite difference semi-discretization* (7.1).

approximations of solutions of the spurious part

(7.56) $$\frac{\partial U}{\partial t} - c\frac{\partial U}{\partial x} = 0$$

of the wave equation (7.16).

The above also reveals an interesting result, namely that the separation of a numerical solution $\{u_n\}$ into its fundamental components $\{p_n\}$ and $\{q_n\}$ corresponds to the separation of its energy $\|u_n\|_2^2$ into $\|p_n\|_2^2$ which propagates forward in the computation, and $\|q_n\|_2^2$ which propagates backward.

7.7. Reflection at a downstream boundary. We are now equipped to solve a number of problems having to do with the behavior of numerical solutions near discontinuities of the computational domain. As an example, we consider the nontrivial problem of describing the spurious reflections that occur near downstream computational boundaries. We examine the finite domain $x \in [0, l]$ on which the model hyperbolic equation (7.2) is approximated with the standard 3-point semi-discretization (7.1).

At the inlet boundary $x = 0$ we let simply

(7.57) $$u_0(t) = U(0, t), \quad \text{imposed},$$

and at the exit boundary $x = l$, the equation is replaced by the 2-point, upwind approximation

(7.58) $$\frac{du_N}{dt} = -c\left(\frac{u_N - u_{N-1}}{h}\right),$$

where

(7.59) $$u_N(t) \simeq U(l, t).$$

One of the effects of this additional (but unavoidable) approximation is to create spurious reflection.

To be specific, consider a numerical solution of $\{p_n\}$ type that has been generated by a (band-limited in $|\Omega h/c| \leq 1$) boundary condition $U(0, t)$, and the subsequent reflected solution (of $\{q_n\}$ type) that is generated when $\{p_n\}$ crosses the exit boundary $x_N = l$.

As before, let $\{\hat{p}_n(\Omega)\}$ and $\{\hat{q}_n(\Omega)\}$ be the corresponding Fourier transforms. In particular, near $x = l$ we may express:

(7.60) $$\begin{aligned}\hat{p}_{N-1} &= \hat{p}_N \hat{E}_1^{-1}, \quad \hat{q}_{N-1} = \hat{q}_N \hat{E}_2^{-1}, \\ \hat{u}_N &= \hat{p}_N + \hat{q}_N, \quad \hat{u}_{N-1} = \hat{p}_{N-1} + \hat{q}_{N-1}.\end{aligned}$$

Using these expressions in taking the Fourier transform of (7.58) results in

(7.61) $$\begin{aligned}i\Omega(\hat{p}_N + \hat{q}_N) &= -\frac{c}{h}(\hat{p}_N + \hat{q}_N - \hat{p}_N \hat{E}_1^{-1} - \hat{q}_N \hat{E}_2^{-1}) \\ &= -\frac{c}{h}(\hat{p}_N + \hat{q}_N + \hat{p}_N \hat{E}_2 + \hat{q}_N \hat{E}_1)\end{aligned}$$

whence, solving for $\hat{q}_N/\hat{p}_N \equiv \rho(\Omega)$ (called the *reflection ratio*), we find

(7.62) $$\rho(\Omega) \equiv \frac{\hat{q}_N}{\hat{p}_N} = -\frac{i(\Omega h/c) + 1 + \hat{E}_2}{i(\Omega h/c) + 1 + \hat{E}_1} = -\frac{1 - \sqrt{1 - (\Omega h/c)^2}}{1 + \sqrt{1 - (\Omega h/c)^2}}.$$

The reflection ratio $\rho(\Omega)$, expresses in the Fourier domain the amount of spurious reflection that occurs at the boundary (see Fig. 7.8).

The incident and reflected solutions at the boundary point may thus be expressed analytically, *exactly*, by the Fourier transforms

(7.63) $$\begin{aligned}p_N(t) &= \int \hat{p}_N(\Omega) e^{i\Omega t} \frac{d\Omega}{2\pi} \\ &= \int \hat{U}(0, \Omega)[\hat{E}_1(\Omega)]^N e^{i\Omega t} \frac{d\Omega}{2\pi}\end{aligned}$$

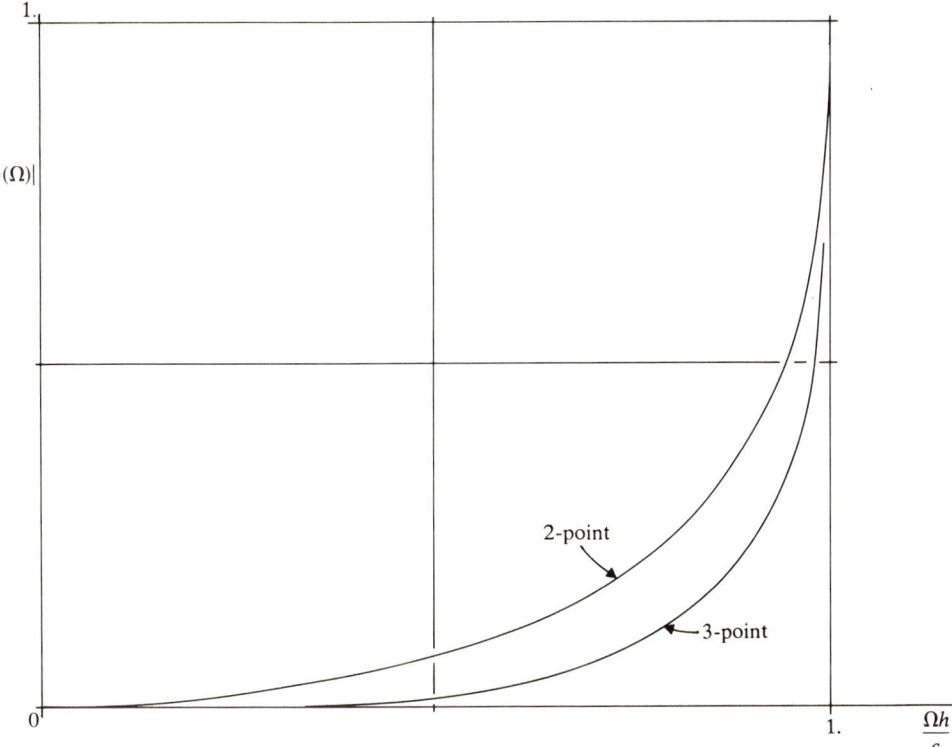

FIG 7.8. *Downstream boundary reflection.*

and

(7.64)
$$q_N(t) = \int \hat{p}_N(\Omega)\rho(\Omega) e^{i\Omega t}\frac{d\Omega}{2\pi}$$
$$= \int \hat{U}(0, \Omega)[\hat{E}_1(\Omega)]^N \rho(\Omega) e^{i\Omega t}\frac{d\Omega}{2\pi}.$$

Note that another form in which (7.62) may be expressed is

(7.65)
$$\rho(\Omega) = -\frac{c - \mathcal{V}(\Omega)}{c + \mathcal{V}(\Omega)},$$

where $\mathcal{V}(\Omega)$ is the group velocity:

(7.66)
$$\mathcal{V}(\Omega) = c\sqrt{1 - \left(\frac{\Omega h}{c}\right)^2}.$$

7.8. Convergence rates. We derive in this section convergence rates for the different components of the numerical solution in the preceding example.

First, we use Fourier analysis to derive the accuracy of the $\{p_n\}$ solution. In any mesh point x_n in $(0, l]$ we have

(7.67) $$\hat{p}_n(\Omega) = \hat{U}(0, \Omega)[\hat{E}_1(\Omega)]^n.$$

By comparison, the Fourier transform of the exact solution at that point is

(7.68) $$\hat{U}(x_n, \Omega) = \hat{U}(0, \Omega) e^{i\Omega nh/c}.$$

Thus, the Fourier transform of the error is

(7.69) $$\hat{\varepsilon}_n(\Omega) = \hat{U}(0, \Omega)[[\hat{E}_1(\Omega)]^n - e^{i\Omega nh/c}].$$

Consider now a fixed point x_A in $(0, l]$, and analyze the convergence rate of $\hat{\varepsilon}_{n_A}$ to zero when $h \to 0$ with $n_A h = x_A = $ constant. Using Taylor's series we find

(7.70) $$\hat{\varepsilon}_{n_A}(\Omega) = \hat{U}(0, \Omega)\left[-\frac{x_A}{h} i \frac{(\Omega h/c)^3}{6} + \text{higher order terms}\right] = O(h^2).$$

This agrees with the fact that the truncation error of (7.1), expressed in classical form is also of second order,

(7.71) $$T_{h,n} \equiv \frac{\partial U_n}{\partial t} + c\left(\frac{U_{n+1} - U_{n-1}}{2h}\right)$$
$$= \frac{c}{6}\left(\frac{\partial^3 U}{\partial x^3}\right)h^2 + \text{higher order terms} = O(h^2).$$

The reflection ratio (7.62) may be expanded in a Taylor series as

(7.72) $$\rho(\Omega) = -\frac{1}{4}\left(\frac{\Omega h}{c}\right)^2 + \text{higher order terms} = O(h^2).$$

Thus, the amplitude of the reflected solution $\{q_n\}$ after passage of $\{p_n\}$ through the exit boundary point $x = l$ is, by (7.64) also of order h^2 (see Fig. 7.9)

(7.73) $$\{q_n\} = O(h^2).$$

It should be noted that the truncation error of the difference formula at the boundary (7.58), is only of first order in h. What we have shown is that despite this lower order of accuracy at the boundary, the global accuracy of the semi-discretization is still of second order:

(7.74) $$\{u_n\} - \{U_n\} = O(h^2).$$

This conclusion verifies a similar result given by Gustafsson (1975), namely that the order of the global accuracy is preserved, in spite of the fact that the accuracy of the approximation at the boundary is one order less.

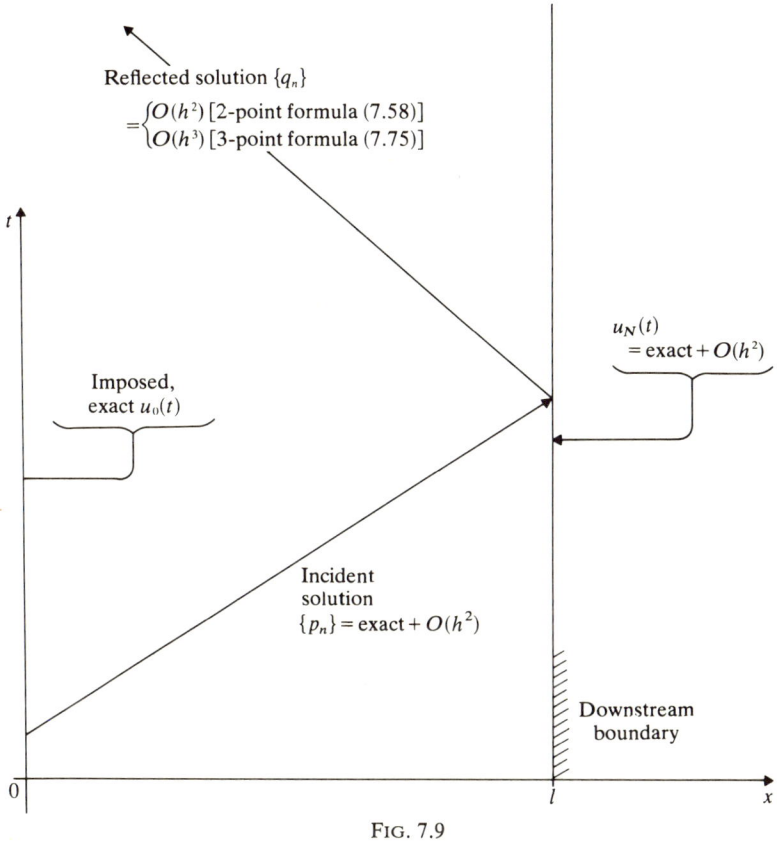

FIG. 7.9

7.9. A 3-point boundary formula. An improved treatment of the exit boundary point $x = x_N$ is the 3-point formula

$$\frac{dU_N}{dt} = -c\left(\frac{3u_N - 4u_{N-1} + u_{N-2}}{2h}\right), \tag{7.75}$$

which has the truncation error

$$\begin{aligned}T_{h,N} &= \frac{\partial U_N}{\partial t} + c\left(\frac{3U_N - 4U_{N-1} + U_{N-2}}{2h}\right) \\ &= -\frac{2}{3}h^2\left(\frac{\partial^3 U}{\partial x^3}\right) + \text{higher order terms} = O(h^2).\end{aligned} \tag{7.76}$$

Proceeding as before (details are left to the reader), we find (see also Fig. 7.8)

$$\rho(\Omega) \equiv \frac{\hat{q}_N(\Omega)}{\hat{p}_N(\Omega)} = \frac{3 + 2i(\Omega h/c) + 4\hat{E}_2 + \hat{E}_2^2}{3 + 2i(\Omega h/c) + 4\hat{E}_1 + \hat{E}_1^2} = O(h^3).$$

Thus, the amplitude of the reflected solution

$$\hat{q}_N(\Omega) = \rho(\Omega)\hat{p}_N(\Omega)$$

converges to zero as $O(h^3)$, as opposed to $O(h^2)$ for the 2-point discretization (7.58). However, the global accuracy of the semi-discretization is still only $O(h^2)$, since the accuracy of $\{p_n\}$ is $O(h^2)$.

Chapter 8

Fourier Analysis and $\mathcal{L}_2$-Norm of the Global Error

8.1. Introduction. What we have done so far is to analyze the accuracy of the time behavior of solutions assumed to be (at least locally) sinusoidal. We have shown in § 2.3 that these sinusoidal solutions can also be interpreted (by letting ω be a parameter rather than a constant) as one component of the Fourier transform of any solution. This gives us the mathematics which are needed to describe the *global* error.

We shall pursue this in some detail in the semi-discrete case. Extension to full discretizations is straightforward.

The set of functions of time

(8.1) $$\{\varepsilon_n(t)\} = \{u_n(t) - U(x_n, t)\}$$

is *the* global error of a semi-discrete approximation. As before, we consider the case where $U(x, t)$ is a solution of the equation $U_t + cU_x = 0$, and $\{u_n(t)\}$ is its approximation.

The discrete $\mathcal{L}_2$-norm of the error

(8.2) $$\|\varepsilon\|_2 = \left(h \sum_n (u_n - U(x_n, t))^2 \right)^{1/2}$$

may be computed by using Parseval's equality. We must first derive the discrete Fourier transform of $\{\varepsilon_n\}$:

(8.3) $$\bar{\varepsilon}(\omega, t) = \bar{u}(\omega, t) - \bar{U}(\omega, t).$$

Here $\bar{U}(\omega, t)$ is the *discrete* Fourier transform of $U(x, t)$, given by (1.56). It is equal to the Fourier transform $\hat{U}(\omega, t)$ if the initial function is band-limited in $|\omega| \le \pi/h$. The discrete Fourier transform of the semi-discrete solution is easily found. With[1]

(8.4) $$\bar{u}(\omega, 0) = \bar{U}(\omega, 0),$$

we find by analytic integration

(8.5) $$\bar{u}(\omega, t) = \bar{U}(\omega, 0) e^{\hat{A}(\omega)t}.$$

On the other hand, for exact solutions

(8.6) $$\bar{U}(\omega, t) = \bar{U}(\omega, 0) e^{-ic\omega t},$$

whence

(8.7) $$\bar{\varepsilon}(\omega, t) = \bar{U}(\omega, 0)(e^{\hat{A}(\omega)t} - e^{-ic\omega t})$$

and, by Parseval's relation (1.62),

(8.8) $$\|\varepsilon\|_2^2 = \int_{-\pi/h}^{\pi/h} |\bar{U}(\omega, 0)|^2 \cdot |e^{\hat{A}(\omega)t} - e^{-ic\omega t}|^2 \frac{d\omega}{2\pi}.$$

8.2. Examples. As an example of application, we analyze the family of semi-discretizations (2.33). We shall be interested in finding what effect the parameter β has upon the capability of the scheme to handle sharp discontinuities or "shocks," as they occur in the equations of compressible gas dynamics. To that effect, we let the initial function $U(x, 0)$ be the step function

(8.9) $$U(x, 0) = \begin{cases} 1 & \text{for } x \geq 0, \\ 0 & \text{for } x < 0. \end{cases}$$

(To be more precise, $U(x, 0)$ is, for $x \geq 0$, $\lim_{\alpha \to 0} e^{-\alpha x}$, leaving $U(\infty, 0) = 0$.)

The discrete Fourier transform $\bar{U}(\omega, 0)$ may be found directly from the definition

(8.10) $$\bar{U}(\omega, 0) = h \sum_{n=0}^{\infty} e^{-i\omega nh} = e^{i\omega h/2} \frac{h/2}{i \sin(\omega h/2)},$$

whence,

(8.11) $$\begin{aligned}\|\varepsilon\|_2^2 &= \int_{-\pi/h}^{\pi/h} \left|\frac{h/2}{\sin(\omega h/2)}\right|^2 \cdot |e^{-i\omega c^*(\omega)t} - e^{-i\omega ct}|^2 \frac{d\omega}{2\pi} \\ &= \int_0^{\pi/h} \left|\frac{h}{\sin(\omega h/2)}\right|^2 \cdot \left|\sin\left(\frac{\omega(c^*(\omega) - c)t}{2}\right)\right|^2 \frac{d\omega}{\pi},\end{aligned}$$

---

[1] We assume that $U(x, 0)$ is band-limited in $[-\pi/h, \pi/h]$, so that no error is introduced by the initial sampling.

where $c^*(\omega)$ is the phase velocity (2.34) of the approximation. If we let $\bar{\omega} = \omega h$ and $\bar{t} = tc/h$, then we find, in dimensionless form, that

$$(8.12) \qquad \frac{1}{h}\|\varepsilon\|_2^2 = \int_0^\pi \left|\frac{\sin(\bar{\omega}(c^*-c)\bar{t}/2c)}{\sin(\bar{\omega}/2)}\right|^2 \frac{d\bar{\omega}}{\pi}.$$

Shown in Fig. 8.1 are values of this function for different values of $\bar{t}$, as a function of $\alpha = \beta/(1-\beta)$ for the family of semi-discretizations (2.33). When $\bar{t} \to \infty$, the least error is obtained with $\alpha = \frac{1}{2}$ (or $\beta = \frac{1}{3}$) which is the value which minimizes the truncation error (in the classical sense, i.e., near $\omega = 0$). By contrast, when $\bar{t} \to 0$, the least error is obtained with $\alpha \simeq 0.8$, which is the value that minimizes the $\mathscr{L}_2$-norm of the truncation error

$$(8.13) \qquad \alpha = \arg\min \|T_h\|_2 \simeq 0.8,$$

$$(8.14) \qquad T_h = \int_{-\pi/h}^{\pi/h} \frac{h/2}{i\sin(\omega h/2)} i\omega(c-c^*) e^{i\omega x} \frac{d\omega}{2\pi}.$$

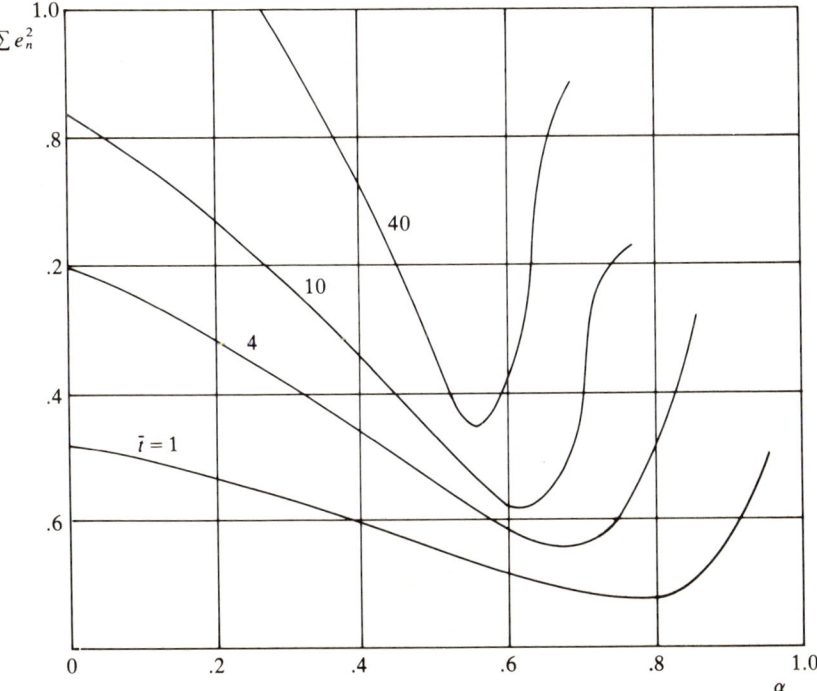

FIG. 8.1. $\mathscr{L}_2$-norm of the global error as a function of the parameter $\alpha = \beta/(1-\beta)$ at different times for the family of semi-discretizations (2.33) and initial conditions (8.9).

As another case we consider an initial step "smoothed" over two increments

(8.15) $$u_n(0) = \begin{cases} 0 & \text{for } n < 0, \\ \frac{1}{2} & \text{for } n = 0, \\ 1 & \text{for } n > 0. \end{cases}$$

Its discrete Fourier transform is

(8.16) $$\bar{u}_n(0) = \frac{h/2}{i \tan(\omega h/2)}.$$

The results are shown in Fig. 8.2.

To illustrate the merits of using the $\mathscr{L}_2$-norm as a measure of the quality of approximations, we show in Fig. 8.3 numerical responses obtained by integrating (2.33) (with $(c\Delta t/h)$ small enough that time discretization errors may be ignored), with initial conditions (8.15) for the time $\bar{t} = 10$. The three cases shown correspond to $\alpha = 0$, .5 and .6, or $\beta = 0$, $\frac{1}{3}$ and $\frac{3}{8}$ respectively, i.e., to the points labeled A, B and C on Fig. 8.2.

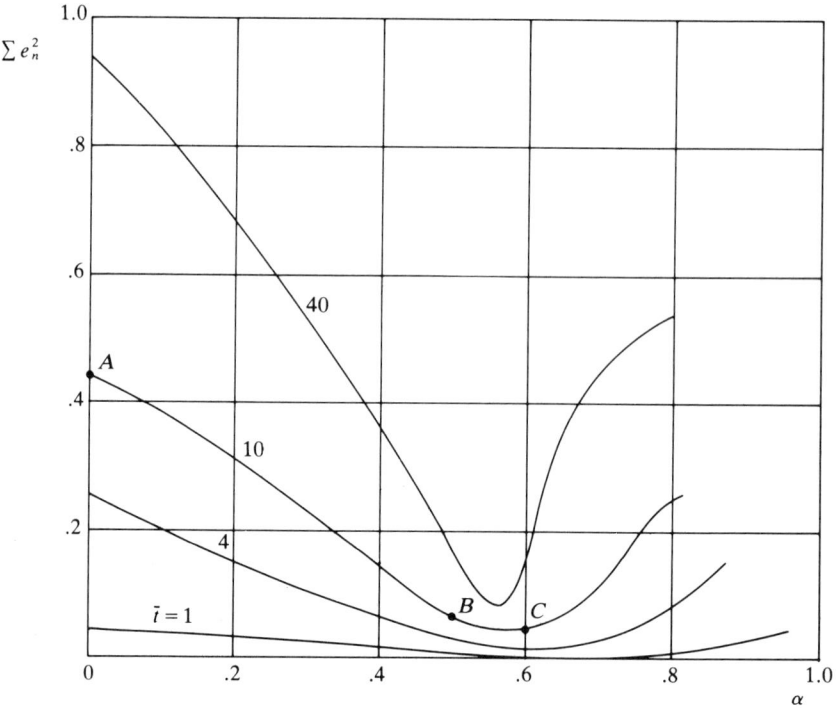

FIG. 8.2. *Same as Fig. 8.1 but with initial conditions* (8.15).

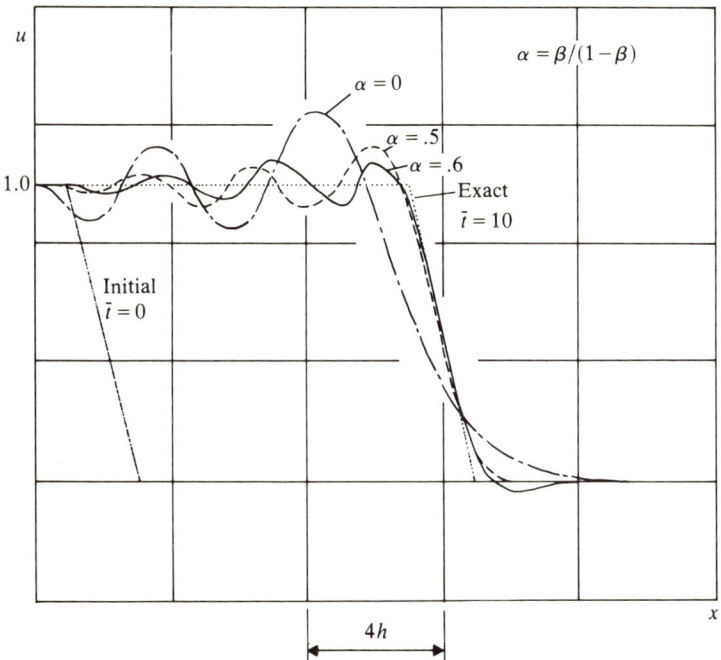

FIG. 8.3. *Numerical solutions of* $\partial U/\partial t + c\, \partial U/\partial x = 0$ *with initial conditions* (8.15) *for different values of* β *in the semi-discretization* (2.33).

8.3. Relation with convergence rates analysis. Expressions of the form (8.8) may serve as a starting point for an analysis of convergence rates when $h \to 0$. For example, for a conservative semi-discretization, the rate at which the $\mathscr{L}_2$-norm of the global error converges to zero is found:

$$(8.17) \quad \lim_{h \to 0} \|\varepsilon\|_2 = \left[\int_{-\pi/h}^{\pi/h} |\bar{U}(\omega, 0) \lim_{h \to 0} (e^{-i\omega c^* t} - e^{-i\omega c t})|^2 \frac{d\omega}{2\pi} \right]^{1/2} = O(h^p),$$

where p is the order of approximation of c by c^* (as in (2.25)). The $\mathscr{L}_2$-norm of the truncation error

$$(8.18) \quad \|T_h\|_2 = \left[\int_{-\pi/h}^{\pi/h} |\bar{U}(\omega, 0) i\omega (c - c^*)|^2 \frac{d\omega}{2\pi} \right]^{1/2} = O(h^p)$$

has the same rate of convergence to zero.

8.4. Asymptotic approximations. The expressions (8.11) and (8.12) are *exact* measures of the discrete $\mathscr{L}_2$-norm of errors. *Approximations* to these expressions may be obtained by ignoring the effect of sampling. For example, to analyze the response of an approximation to a step initial condition, we may approximate

$$\bar{U}(\omega, 0) = -\frac{ih/2}{\sin(\omega h/2)}$$

by $\hat{U}(\omega, 0) = 1/i\omega$, whence, using complete rather than discrete Fourier transforms we have

(8.19) $$\|\varepsilon\|_2^2 \simeq \int_{-\infty}^{\infty} \left|\frac{1}{\omega}\right|^2 \cdot \left|2\sin\left(\frac{\omega(c^*(\omega)-c)t}{2}\right)\right|^2 \frac{d\omega}{2\pi}.$$

The approximation so obtained is smaller than (8.11), (since a sum of squares is less than the square of the sum). However, the difference is not very great, and such asymptotic approximations have occasionally been used to evaluate discrete approximations of hyperbolic equations, e.g., by Wesseling (1972), (1973) and by Miranker (1971).

Chapter 9

Spectral Methods

9.1. Introduction. Other than finite difference, finite element and spline Galerkin methods used to semi-discretize hyperbolic equations, attention has been given in recent years to methods using *trigonometric interpolation*. A natural use of these methods occurs with problems having periodic boundary conditions. Specifically, consider as a model the equation

(9.1) $$\frac{\partial U}{\partial t} + c \frac{\partial U}{\partial x} = G(x, t)$$

over $x \in [0, l)$ with boundary condition

(9.2) $$U(0, t) = U(l, t).$$

Periodic boundary conditions of this kind occur for instance in mathematical models of the global atmosphere and the methods we describe next have indeed been developed and used to a large extent by geophysical fluid dynamicists.

Trigonometric approximation. Any function $U(x)$ defined over $[0, l)$, which is square summable over that domain, may be expressed in Fourier series form as

(9.3) $$U(x) = \frac{1}{l} \sum_k \hat{U}_k e^{2ik\pi x/l},$$

where the $\hat{U}_k$ are given by the transform

(9.4) $$\hat{U}_k = \int_0^l U(x) e^{-2ik\pi x/l} dx.$$

An *analytic* solution to the problem (9.1)–(9.2) is of course easily expressed in the frequency (or ^) domain. Since

(9.5) $$\frac{\partial U}{\partial x} = \frac{1}{l} \sum_k \frac{2ik\pi}{l} \hat{U}_k e^{2ik\pi x/l},$$

it follows that an expression of (9.1) is

$$\text{(9.6)} \qquad \frac{1}{l} \sum_k \left[\frac{d\hat{U}_k}{dt} + c\left(\frac{2ik\pi}{l}\right) \hat{U}_k = \hat{g}_k \right] e^{2ik\pi x/l},$$

where the $\{\hat{g}_k\}$ are the Fourier transforms of the function $G(x, t)$. Using the orthogonality property of trigonometric terms results in the (infinite) system of ordinary differential equations,

$$\text{(9.7)} \qquad \frac{d\hat{U}_k}{dt} = -c\frac{2ik\pi}{l} \hat{U}_k + \hat{g}_k, \qquad k = \cdots, -2, -1, 0, 1, 2, \cdots.$$

Approximations are obtained when $U(x, t)$ is replaced by a series such as (9.3), but with k restricted to within a finite range of values, say $[-N, N]$, i.e., $U(x, t)$ is approximated by a *truncated* Fourier series. Using this concept leads to two types of feasible approximations:

(i) a truncated Fourier series method in which time integration is done in the Fourier coefficients $\hat{U}_k(t)$ and all calculations take place in the frequency domain;

(ii) a collocation method in which time integration is done by expressing the solution in physical space (rather than the frequency domain) and trigonometric interpolation is used solely with the intent of approximating the partial derivative $\partial U/\partial x$ at a finite number of solution (or collocation) points $\{x_n; n = 0, 1, 2, \cdots\}$.

Both of these methods are explained below.

9.2. Truncated Fourier series method. We must assume here that the velocity c in (9.1) is independent of x. This assumption is decisive for the following. Let $h = l/2N$ and

$$\text{(9.8)} \qquad x_n = nh, \qquad u_n \simeq U(x_n, t), \qquad n = 0, 1, 2, \cdots, 2N-1,$$

be a set of discrete solution points. Between those points, we express the approximate solution as a truncated Fourier series

$$\text{(9.9)} \qquad U(x, t) \simeq u(x, t) = \frac{1}{l} \sum_{k=-N}^{N} \bar{u}_k(t) e^{2ik\pi x/l},$$

which is the trigonometric interpolant between the point values (9.8) of the numerical solution. The relation giving the Fourier coefficients $\{\bar{u}_k(t)\}$ is the discrete transform

$$\text{(9.10)} \qquad \bar{u}_k = h \sum_{n=0}^{2N-1} u_n e^{-2ik\pi x_n/l}.$$

Expressing satisfaction of the differential equation (9.1) by the approximate solution results in

$$\sum_k \left(\frac{d\bar{u}_k}{dt} + c \frac{2ik\pi}{l} \bar{u}_k \right) e^{2ik\pi x/l} = G(x, t), \tag{9.11}$$

and we may recast this equation in a component-by-component form as

$$\frac{d\bar{u}_k}{dt} + c \frac{2ik\pi}{l} \bar{u}_k = \bar{g}_k(t), \qquad k = -N, \cdots, 0, 1, 2, \cdots, N. \tag{9.12}$$

This represents a total of $2N$ real equations (i.e., a number equal to the number of solution points in physical space).

Here the $\{\bar{g}_k(t)\}$ are the discrete Fourier components of G, i.e.,

$$\bar{g}_k = h \sum_{n=0}^{2N-1} G(x_n, t) e^{-2ik\pi x_n/l}. \tag{9.13}$$

The semi-discrete approximation (9.12) may also be arrived at formally by applying the Galerkin procedure: Define the equation residual

$$\mathcal{R}(x, t) \equiv \frac{\partial u}{\partial t} + c \frac{\partial u}{\partial x} - G = \sum_k \left[\left(\frac{d\bar{u}_k}{dt} + c \frac{2ik\pi}{l} \bar{u}_k \right) - \bar{g}_k \right] e^{2ik\pi x_n/l}. \tag{9.14}$$

The discrete expression of the Galerkin conditions

$$\langle e^{-2ik\pi x_n/l}, \mathcal{R} \rangle \equiv h \sum_{n=0}^{2N-1} e^{-2ik\pi x_n/l} \mathcal{R} = 0,$$
$$k = -N, -N+1, \cdots, 0, 1, \cdots, N \tag{9.15}$$

results precisely in the $2N$ equations (9.12). The solution is then obtained by integrating the ordinary differential equations (9.12) with respect to time for the Fourier coefficients $\{\bar{u}_k(t)\}$ using any one of the marching methods which are suitable for hyperbolic equations (explicit or implicit).

The entire calculation takes place in the frequency domain. But as soon as the problem becomes nonlinear and/or the coefficients of the equation (such as c) become space dependent, then the simplicity of this method disappears.

9.3. Analysis.

Error analysis. There are two sources of error in the preceding procedure:
(i) the approximation of $G(x, t)$ by a *truncated* Fourier series;
(ii) the approximation of the initial function $U(x, 0)$ by a *truncated* Fourier series.

No error is introduced otherwise in the replacement of the partial derivative $\partial U/\partial x$ by the truncated series contained in (9.11). If both $U(x, 0)$ and $G(x, t)$ are band-limited in $[-2N\pi/l, 2N\pi/l] = [-\pi/h, \pi/h]$ then the semi-discretization (9.12) is exact.

Time marching—stability. The integration with respect to time is that of the system of complex ordinary differential equations, (9.12), i.e., a system of pairs of real ordinary differential equations. The eigenvalues

$$\frac{2ick\pi}{l}, \quad k = -N, -N+1, \cdots, 1, 2, \cdots, N \tag{9.16}$$

play here the role of the spectral function, and the condition of numerical stability which replaces (4.14) is

$$\max_{k} \left|\frac{2c\,\Delta t k\pi}{l}\right| = \left|\frac{2c\,\Delta t N\pi}{l}\right| \leq S_I, \tag{9.17}$$

where the S_I are those given in Table 4.2. For example, for the leapfrog method, $S_I = 1$ and the condition of numerical stability is

$$\left|\frac{2c\,\Delta t N\pi}{l}\right| = \left|\frac{c\,\Delta t\pi}{h}\right| \leq 1$$

or

$$\left|\frac{c\,\Delta t}{h}\right| \leq \frac{1}{\pi}. \tag{9.18}$$

9.4. Fourier "collocation" method. The second type of trigonometric approximation relies on the expression of the solution in the form of a truncated Fourier series *only as an intermediate step to accurately computing the spatial derivatives.* Semi-discrete equations are expressed in terms of time derivatives of the *nodal values* $\{du_n/dt\}$, which are integrated from time step to time step in the $\{u_n^j\}$. The name Fourier "collocation" method is used by Orszag to describe this method, although this is *not* a collocation method in the strict sense.

Early publications describing this class of algorithms are those of Orszag (1971), Kreiss and Oliger (1972) and Fornberg (1972). Methods of this kind are being actively investigated in the context of geophysical fluid dynamics under the general heading of *"spectral"* and *"pseudo-spectral"* methods.

The Fourier "collocation" method may be described by using the simple model equation

$$\frac{\partial U}{\partial t} + c(x, t)\frac{\partial U}{\partial x} = G(x, t) \quad \text{in } x \in [0, l). \tag{9.19}$$

with periodic boundary condition

$$U(0, t) = U(l, t). \tag{9.20}$$

Note that in contrast with (9.1), c may now be x-dependent. A grid of equidistant solution points is again selected:

(9.21) $$x_n = nh, \quad n = 0, 1, 2, \cdots, 2N-1, \quad h = \frac{l}{2N}.$$

To compute the $\{(\partial u/\partial x)_n\}$ in those grid points, the solution is approximated by the truncated Fourier series

(9.22) $$U(x, t) \simeq u(x, t) = \frac{1}{l} \sum_{k=-N}^{N} \bar{u}_k(t) e^{2ik\pi x/l}$$

and the Fourier coefficients $\bar{u}_k$ are related to the nodal values $\{u_n\}$ by the usual discrete Fourier transform:

(9.23) $$\bar{u}_k(t) = h \sum_{n=0}^{2N-1} u_n(t) e^{-2ik\pi x_n/l}.$$

The approximation of the derivatives at the mesh points are taken to be the analytic derivatives of (9.22), i.e.,

(9.24) $$\left(\frac{\partial u}{\partial x}\right)_n = \frac{1}{l} \sum_k \bar{u}_k(t) \frac{2ik\pi}{l} e^{2ik\pi x_n/l}.$$

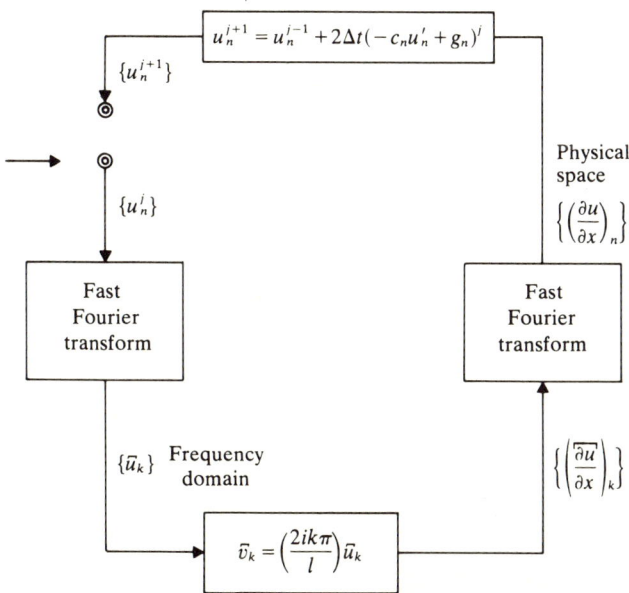

FIG. 9.1. *Computing block diagram for the pseudo-spectral method with leapfrog time marching.*

The semi-discretization of (9.19) is then simply the set of ordinary differential equations in physical $\{u_n\}$ space

$$(9.25) \qquad \frac{du_n}{dt} = -c(x_n, t)\left(\frac{\partial u}{\partial x}\right)_n + G(x_n, t),$$

where the $(\partial u/\partial x)_n$ are to be computed by (9.24). These are then integrated in time by any one of the usual marching methods which are suitable for hyperbolic equations.

Most implementations of the Fourier collocation method use the fast Fourier transform algorithm to effect the transforms (9.23) and (9.24). The procedure for solving a hyperbolic equation such as (9.19) can be illustrated by a diagram such as in Fig. 9.1.

Note that if c were not a function of x, then the two semi-discretizations (9.24) and (9.12) would be equivalent, and either Fourier method could be used. But (as is the case with most real life problems) if c does depend on x then the first Fourier method can no longer be applied (at least not without significant complications) while the second method (leading to (9.25)) remains applicable.

Chapter 10

Equations in Two Dimensions: Anisotropy

10.1. The advection equation in two dimensions. Another type of error occurs in the numerical approximation of hyperbolic equations in two dimensions. In addition to being dependent upon frequency, the velocity of propagation of numerical sinusoidal solutions is also dependent upon direction. That is, in addition to numerical dispersion, we also have numerical *anisotropy*.

To illustrate this, consider the advection equation in two dimensions

(10.1) $$\frac{\partial U}{\partial t} + c_x \frac{\partial U}{\partial x} + c_y \frac{\partial U}{\partial y} = 0, \quad U = U(x, y, t)$$

where

$$c_x = c \cos \alpha, \quad c_y = c \sin \alpha$$

are velocity components in the x and y directions respectively. We may rewrite this equation as

(10.1') $$\frac{\partial U}{\partial t} + \mathbf{c}^T \cdot \nabla U = 0,$$

where $\mathbf{c}$ is the velocity vector in the plane. This notation underscores the independence of the equation from the system of coordinates used. While we may write in cartesian coordinates

(10.2) $$\mathbf{c} = \begin{pmatrix} c_x \\ c_y \end{pmatrix}, \quad \nabla \cdot = \begin{pmatrix} \frac{\partial}{\partial x} \\ \frac{\partial}{\partial y} \end{pmatrix},$$

the notation (10.1') also holds in any other coordinate system in the plane.

10.2. Anisotropy of the approximation on a square grid. A simple semi-discretization of the equation is obtained with Cartesian coordinates on a square grid,

$$x_m = mh, \qquad y_n = nh,$$

where we may write

(10.3) $$\frac{du_{m,n}}{dt} = -c_x\left(\frac{u_{m+1,n} - u_{m-1,n}}{2h}\right) - c_y\left(\frac{u_{m,n+1} - u_{m,n-1}}{2h}\right)$$

with $u_{m,n}(t) \simeq U(x_m, y_n, t)$. The two-dimensional equivalent of the sinusoidal trial solutions used before are sinusoidal plane (or rather line) solutions, which may be written as

(10.4) $$U_\omega(x, y, t) = a_\omega(t)\, e^{i\omega(x\cos\alpha + y\sin\alpha)}$$
$$= a_\omega(t)\, e^{i\omega(\mathbf{r}^T \cdot \mathbf{1}_\alpha)} = a_\omega(t) \cdot e^{i(\omega_x x + \omega_y y)}.$$

Here **r** is the position vector in the (x, y)-plane,

(10.5) $$\mathbf{r} = \begin{pmatrix} x \\ y \end{pmatrix}.$$

$\mathbf{1}_\alpha$, the directional vector of the wave, is chosen parallel to the velocity vector,

(10.6) $$\mathbf{1}_\alpha = \begin{pmatrix} \cos\alpha \\ \sin\alpha \end{pmatrix}, \qquad \mathbf{c} = c \cdot \mathbf{1}_\alpha = \begin{pmatrix} c\cos\alpha \\ c\sin\alpha \end{pmatrix}$$

and

$$\omega_x = \omega\cos\alpha, \quad \omega_y = \omega\sin\alpha, \quad \omega = (\omega_x^2 + \omega_y^2)^{1/2}.$$

The wavelength (measured in the $\mathbf{1}_\alpha$-direction) is

(10.7) $$\lambda = \frac{2\pi}{\omega}.$$

Upon insertion of (10.4) into (10.1), we find

(10.8) $$\frac{da_\omega}{dt} = -ic\omega(\cos^2\alpha + \sin^2\alpha)a_\omega = -ic\omega a_\omega,$$

whence the equation of *exact* sinusoidal solutions in the plane,

(10.9) $$U_\omega(x, y, t) = a_\omega(0)\, e^{i\omega(x\cos\alpha + y\sin\alpha - ct)}.$$

Constant phase lines

(10.10) $$x\cos\alpha + y\sin\alpha - ct = \text{constant}$$

move as expected at the velocity c in the $\mathbf{1}_\alpha$ direction (see Fig. 10.1).

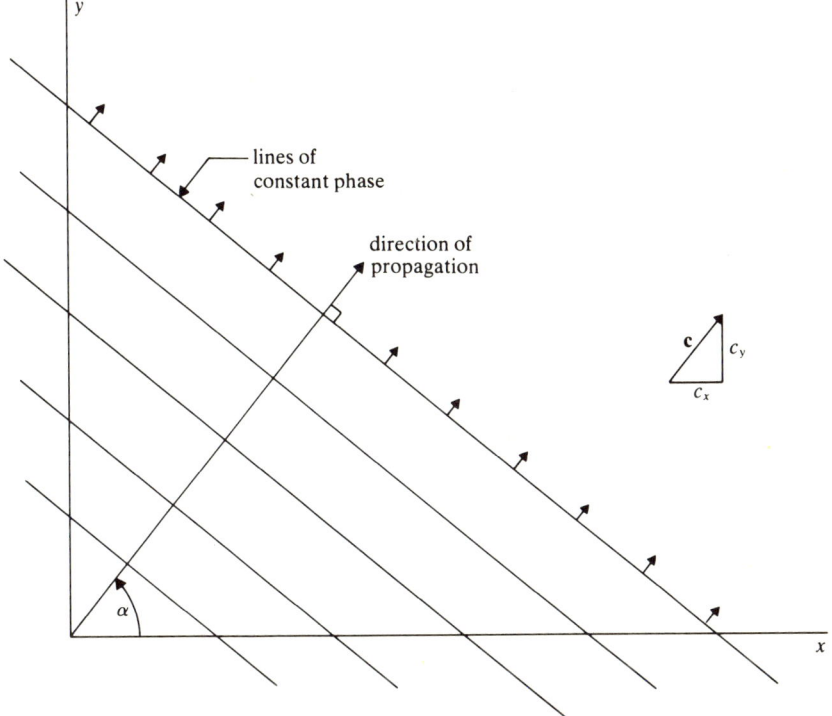

FIG. 10.1. *Illustration of sinusoidal wave lines in the (x, y)-plane.*

Sinusoidal *numerical* solutions are of the form

(10.11) $$u_{m,n}(t) = a_\omega(t) \, e^{i\omega(x_m \cos\alpha + y_n \sin\alpha)}.$$

Their propagation properties are obtained by insertion into (10.3), producing

(10.12) $$\frac{da_\omega}{dt} = -\left(ic_x \frac{\sin(\omega_x h)}{h} + ic_y \frac{\sin(\omega_y h)}{h}\right) a_\omega$$

$$= -\frac{ic}{h}(\cos\alpha \, \sin(\omega_x h) + \sin\alpha \, \sin(\omega_y h)) a_\omega.$$

We find by analytic integration

(10.13) $$a_\omega(t) = a_\omega(0) \, e^{-i\omega c^* t},$$

where

(10.14) $$c^* = \frac{c}{\omega h}(\cos\alpha \, \sin(\omega_x h) + \sin\alpha \, \sin(\omega_y h))$$

is the numerical phase velocity.

Numerical sinusoidal solutions are thus

(10.15) $$u_{m,n}(t) = a_\omega(0)\, e^{i\omega(x_m \cos\alpha + y_n \sin\alpha - c^* t)}.$$

Constant phase lines are expressed by the equation

(10.16) $$x \cos\alpha + y \sin\alpha - c^* t = \text{constant}.$$

They are perpendicular to $\mathbf{1}_\alpha$ in the (x, y)-plane, moving in the $\mathbf{1}_\alpha$ direction at the velocity c^*.

As in the one-dimensional case, the phase velocity is a function of the frequency ω (or wavelength $\lambda = 2\pi/\omega$).

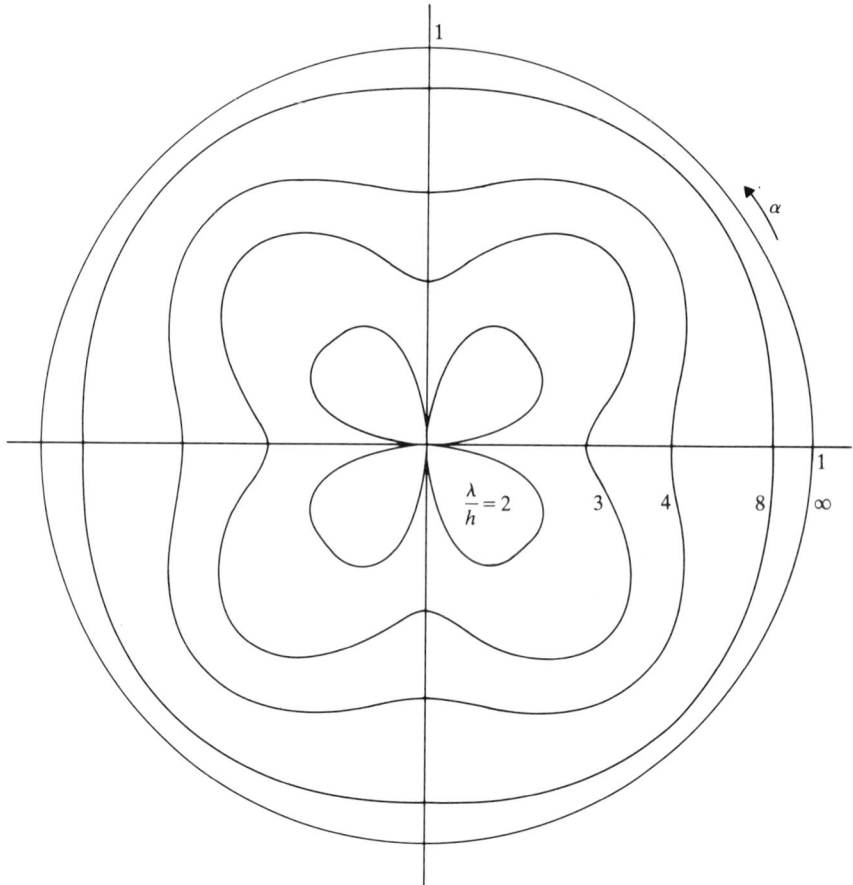

FIG. 10.2. *Polar diagram of the normalized phase velocity c^*/c as a function of wavelength λ and direction α for the semi-discretization* (10.3) *of the advection equation in two dimensions. Note the preferential direction of propagation along the lines at $\pm 45°$.*

EQUATIONS IN TWO DIMENSIONS: ANISOTROPY 119

But in addition, c^* (see (10.14)) is also a function of the direction of propagation

(10.17) $$c^* = c^*(\omega, \alpha).$$

It is in this directional dependence of the phase velocity that the anisotropy of the numerical approximation lies.

A convenient way to describe $c^*(\omega, \alpha)$ graphically is by the use of a polar diagram. Shown in Fig. 10.2 is the numerical phase velocity (10.14) for different values of the wavelength $\lambda = 2\pi/\omega$.

10.3. Analysis of other explicit formulae. Other explicit semi-discretizations of the advection equation (10.1) may be derived. We consider first the 9-point formula obtained by expressing a more general approximation to the spatial derivatives. With operators $\boldsymbol{E}_x \cdot$, $\boldsymbol{E}_y \cdot$, $\boldsymbol{D}_x \cdot$, $\boldsymbol{D}_y \cdot$ defined as

$$\boldsymbol{E}_x \cdot u_{m,n} = u_{m+1,n}, \qquad \boldsymbol{E}_y \cdot u_{m,n} = u_{m,n+1},$$

$$\boldsymbol{D}_x \cdot u_{m,n} = (\boldsymbol{E}_x - \boldsymbol{E}_x^{-1}) \cdot u_{m,n} = u_{m+1,n} - u_{m-1,n},$$

$$\boldsymbol{D}_y \cdot u_{m,n} = (\boldsymbol{E}_y - \boldsymbol{E}_y^{-1}) \cdot u_{m,n} = u_{m,n+1} - u_{m,n-1},$$

a general 9-point approximation of (10.1) is

(10.18)
$$\frac{du_{m,n}}{dt} = -c_x \left[1 - \beta + \frac{\beta}{2}(\boldsymbol{E}_y + \boldsymbol{E}_y^{-1}) \right] \cdot \frac{\boldsymbol{D}_x}{2h} \cdot u_{m,n}$$
$$- c_y \left[1 - \beta + \frac{\beta}{2}(\boldsymbol{E}_x + \boldsymbol{E}_x^{-1}) \right] \cdot \frac{\boldsymbol{D}_y}{2h} \cdot u_{m,n},$$

where β is a free parameter. That is, all nine points in the $(m \pm 1, n \pm 1)$ square are invoked to approximate both $\partial U/\partial x$ and $\partial U/\partial y$ (see Fig. 10.3).

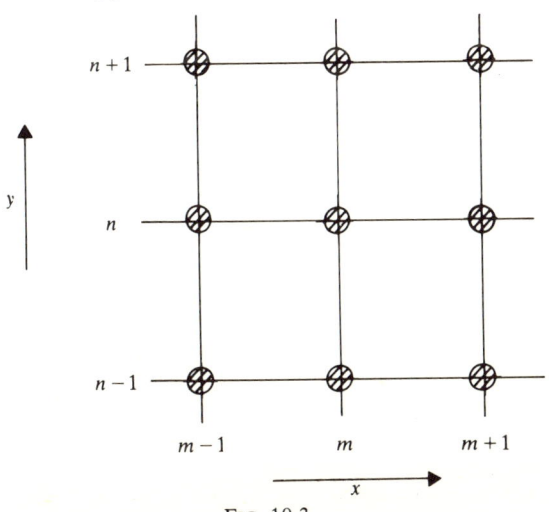

FIG. 10.3

As before, these approximations are conservative and the discretization error is entirely described by the expression of the numerical phase velocity. Repeating the previous mathematics, we now find

(10.19)
$$c^*(\omega, \alpha) = c\left[\cos^2\alpha((1-\beta)+\beta\cos(\omega_y h))\left(\frac{\sin(\omega_x h)}{\omega_x h}\right)\right.$$
$$\left.+\sin^2\alpha((1-\beta)+\beta\cos(\omega_x h))\left(\frac{\sin(\omega_y h)}{\omega_y h}\right)\right].$$

This function is shown for different values of β in Figs. 10.4a–10.4d.

10.4. Implicit approximations. A family of approximations for the two-dimensional advection equation which generalizes the *implicit* semi-discretization (2.33) is that described by the formula (notation is the same as in (10.18))

(10.20)
$$\left[1-\beta+\frac{\beta}{2}(E_x+E_x^{-1})\right]\cdot\left[1-\beta+\frac{\beta}{2}(E_y+E_y^{-1})\right]\cdot\frac{du_{m,n}}{dt}$$
$$=-c_x\left[1-\beta+\frac{\beta}{2}(E_y+E_y^{-1})\right]\cdot\frac{D_x}{2h}u_{m,n}$$
$$-c_y\left[1-\beta+\frac{\beta}{2}(E_x+E_x^{-1})\right]\cdot\frac{D_y}{2h}u_{m,n}.$$

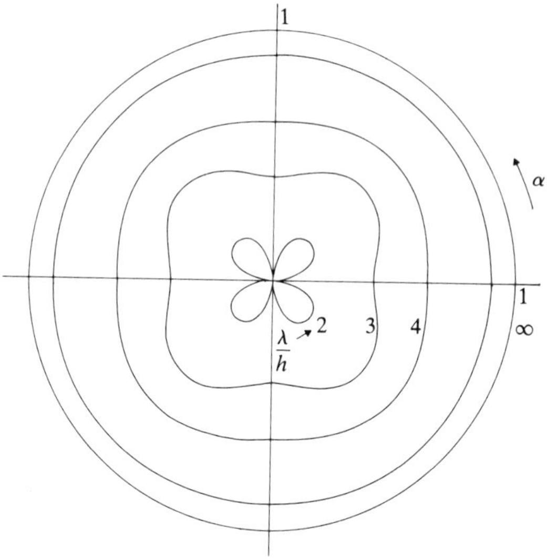

FIG. 10.4a. *Polar diagram of the normalized phase velocity c^*/c as a function of wavelength λ and direction α for the semi-discretization (10.18) of the advection equation in two dimensions. This semi-discretization uses a 9-point approximation to the spatial derivatives. Here $\beta = .25$.*

EQUATIONS IN TWO DIMENSIONS: ANISOTROPY 121

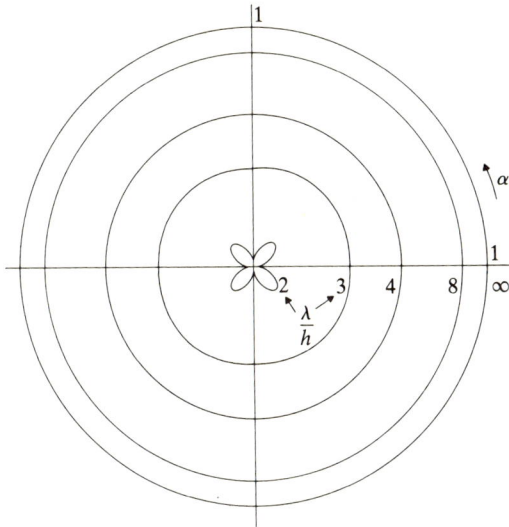

FIG. 10.4b. *Same as Fig. 10.4a but with $\beta = .4$. Note that except for short wavelengths the propagation is almost isotropic.*

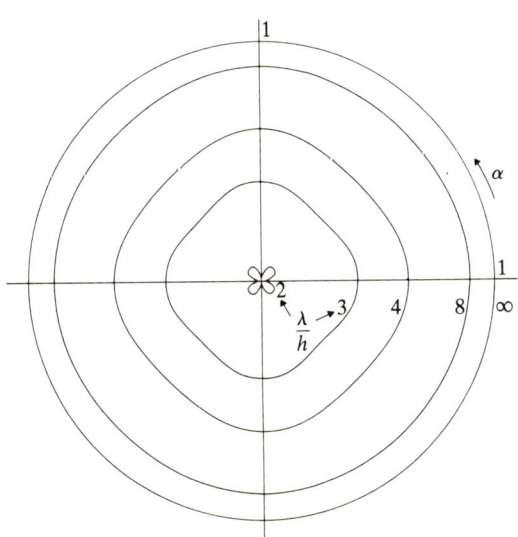

FIG. 10.4c. *Same as Fig. 10.4a but with $\beta = .5$.*

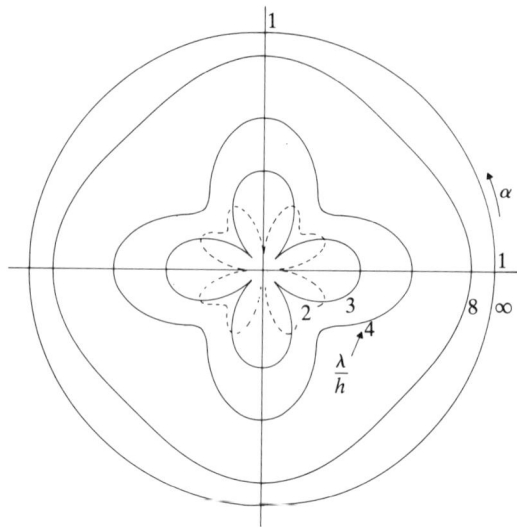

FIG. 10.4d. *Same as Fig. 10.4a but with $\beta = 1$. Note that the directions of preferential propagation are now the axes $0°$ and $90°$ and that c^* for $\lambda/h = 2$ is negative.*

When $\beta = \frac{1}{3}$, this is the expression found with the Galerkin method using bilinear finite elements on squares. The expression of the numerical velocity is easily found (see Fig. 10.5):

(10.21)
$$c^* = c\left[\frac{\cos^2 \alpha}{1-\beta+\beta \cos(\omega_x h)} \frac{\sin(\omega_x h)}{\omega_x h} + \frac{\sin^2 \alpha}{1-\beta+\beta \cos(\omega_y h)} \frac{\sin(\omega_y h)}{\omega_y h}\right].$$

10.5. The wave equation in the plane. Consider now the wave equation in the plane

(10.22a)
$$\frac{\partial^2 U}{\partial t^2} = c^2 \cdot \nabla^2 U.$$

In Cartesian coordinates (x, y) this equation becomes

(10.22b)
$$\frac{\partial^2 U}{\partial t^2} = c^2 \left(\frac{\partial^2 U}{\partial x^2} + \frac{\partial^2 U}{\partial y^2}\right).$$

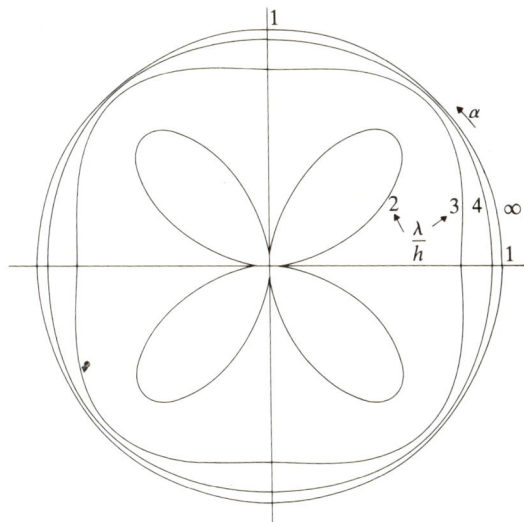

FIG. 10.5. *Polar diagram showing the normalized phase velocity c^*/c as a function of wavelength λ and direction α for the semi-discretization* (10.20) *with $\beta = \frac{1}{3}$. This is identical to the semi-discretization obtained with the finite element/Galerkin method with bilinear elements on the square.*

A simple finite difference semi-discretization of this form over a square mesh is

(10.23)
$$\frac{d^2 u_{m,n}}{dt^2} = \frac{c^2}{h^2}(u_{m-1,n} + u_{m+1,n} + u_{m,n-1} + u_{m,n+1} - 4u_{m,n})$$
$$= \frac{c^2}{h^2}(\delta_x^2 + \delta_y^2) \cdot u_{m,n},$$

where

$$\delta_x^2 u_{m,n} \equiv u_{m-1,n} - 2u_{m,n} + u_{m+1,n},$$
$$\delta_y^2 u_{m,n} \equiv u_{m,n-1} - 2u_{m,n} + u_{m,n+1}.$$

For convenience, we shall also use the notation

(10.23')
$$\frac{d^2 u_{m,n}}{dt^2} = c^2 \diamond \cdot u_{m,n},$$

where

$$\diamond \cdot \equiv \frac{\delta_x^2 + \delta_y^2}{h^2}.$$

is a 5-point discrete approximation to the Laplacian $\nabla^2 \cdot$ (see Fig. 10.6).

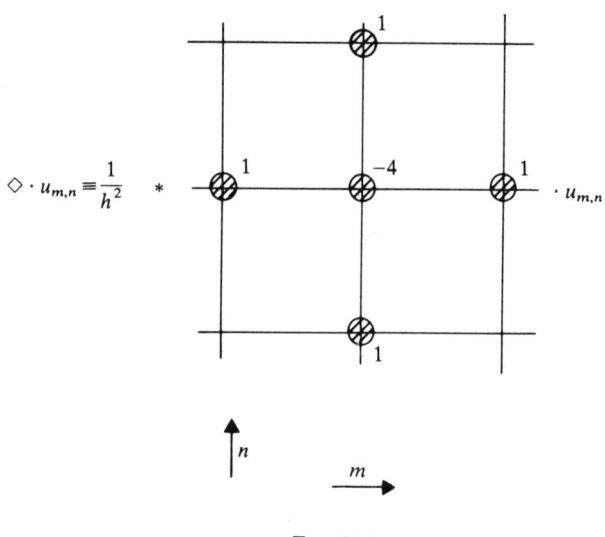

FIG. 10.6

Time integration may be achieved as with the one-dimensional wave equation. For example, the replacement of the time derivative by a standard 3-point finite difference approximation gives

(10.24)
$$u_{m,n}^{j+1} = -u_{m,n}^{j-1} + 2u_{m,n}^{j} + \left(\frac{c\Delta t}{h}\right)^2 (\delta_x^2 + \delta_y^2) \cdot u_{m,n}^{j}$$
$$= -u_{m,n}^{j-1} + 2u_{m,n}^{j} + (c\Delta t)^2 \diamond \cdot u_{m,n}^{j},$$

or one may use the other methods described in § 4.5.

As underscored by the notation (10.22a), the wave equation is *isotropic*: properties of wave propagation are independent of the direction of propagation. A weakness of finite-difference discretizations of the form (10.24) is that isotropy is lost. As we shall see, the properties of numerical wave propagation become a function of the direction.[1]

To demonstrate this, we shall analyze the error in sinusoidal wave propagation due to the semi-discretization (10.23). (This is equivalent to analyzing the error when the Courant number $c\Delta t/h \to 0$.)

Exact sinusoidal solutions. By substitution of (10.4) into (10.22b) we find

(10.25)
$$\frac{d^2 a_\omega}{dt^2} e^{i(\omega_x x + \omega_y y)} = -a_\omega c^2 (\omega_x^2 + \omega_y^2) e^{i(\omega_x x + \omega_y y)}$$

---

[1] See Birkhoff and Dougalis (1975).

or (since $\omega_x^2 + \omega_y^2 = \omega^2$)

(10.26) $$\frac{d^2 a_\omega}{dt^2} = -(\omega c)^2 a_\omega.$$

Solving analytically,

(10.27) $$a_\omega(t) = a_\omega(0) e^{\pm i\omega c t},$$

we may thus express $U(x, y, t)$ as

(10.28) $$U_\omega(x, y, t) = a_\omega(0) e^{i\omega(x \cos \alpha + y \sin \alpha \pm ct)}.$$

Constant phase,

(10.29) $$x \cos \alpha + y \sin \alpha \pm ct = \text{constant}$$

is the equation of two straight lines which are perpendicular to the direction 1_α and which move at the velocity $\pm c$ in opposite directions. That this velocity is independent of α is an expression of the isotropy property of the wave equation.

Numerical sinusoidal solutions. Numerical sinusoidal solutions are, as before,

(10.30) $$u_{m,n}(t) = a_\omega(t) e^{i\omega(x_m \cos \alpha + y_n \sin \alpha)}.$$

Substitution into (10.23) produces, after elimination of common terms,

(10.31) $$\begin{aligned}\frac{d^2 a_\omega}{dt^2} &= c^2 \left[\frac{(2 \cos(\omega_x h) - 2) + (2 \cos(\omega_y h) - 2)}{h^2} \right] a_\omega \\ &= -(\omega c)^2 \left[\cos^2 \alpha \left(\frac{\sin(\omega_x h/2)}{\omega_x h/2} \right)^2 + \sin^2 \alpha \left(\frac{\sin(\omega_y h/2)}{\omega_y h/2} \right)^2 \right] a_\omega.\end{aligned}$$

The solution is

$$a_\omega(t) = a_\omega(0) e^{\pm i\omega c^* t},$$

where

(10.32) $$\begin{aligned}c^* &= c \left[\cos^2 \alpha \left(\frac{\sin(\omega_x h/2)}{\omega_x h/2} \right)^2 + \sin^2 \alpha \left(\frac{\sin(\omega_y h/2)}{\omega_y h/2} \right)^2 \right]^{1/2} \\ &= c \frac{\sqrt{2}}{\omega h} (2 - \cos(\omega_x h) - \cos(\omega_y h))^{1/2} \\ &= c^*(\alpha, \omega)\end{aligned}$$

is the numerical phase velocity of sinusoidal plane waves. Thus

(10.33) $$u_{m,n} = a_\omega(0) e^{i\omega(x_m \cos \alpha + y_n \sin \alpha \pm c^*(\alpha, \omega) t)}.$$

126   CHAPTER 10

As before, it is in the α dependence of $c^*(\omega, \alpha)$ that the anisotropy of the numerical approximation lies.

In Fig. 10.7 we have plotted $c^*(\omega, \alpha)$ in polar form as a function of its two parameters. We may observe that for a given wavelength, velocities are higher (i.e., more accurate) in directions which are at ±45° with respect to the x and y axes.

Another simple 5-point semi-discretization is obtained by a rotation of the stencil of (10.23) by 45°, and a replacement of h by $\sqrt{2}\, h$, thus resulting in

(10.34)
$$\frac{d^2 u_{m,n}}{dt^2} = \frac{c^2}{2h^2}(u_{m+1,n+1} + u_{m+1,n-1} + u_{m-1,n+1} + u_{m-1,n-1} - 4u_{m,n})$$
$$\equiv c^2 \square \cdot u_{m,n},$$

where the operator $\square \cdot$ is defined as (see Fig. 10.8)

(10.35)
$$\square \cdot \equiv \frac{1}{2h^2}(\boldsymbol{E}_x \boldsymbol{E}_y \cdot + \boldsymbol{E}_x \boldsymbol{E}_y^{-1} \cdot + \boldsymbol{E}_x^{-1} \boldsymbol{E}_y \cdot + \boldsymbol{E}_x^{-1} \boldsymbol{E}_y^{-1} \cdot - 4)$$
$$= \frac{1}{2h^2}((\boldsymbol{E}_x + \boldsymbol{E}_x^{-1})(\boldsymbol{E}_y + \boldsymbol{E}_y^{-1}) \cdot - 4).$$

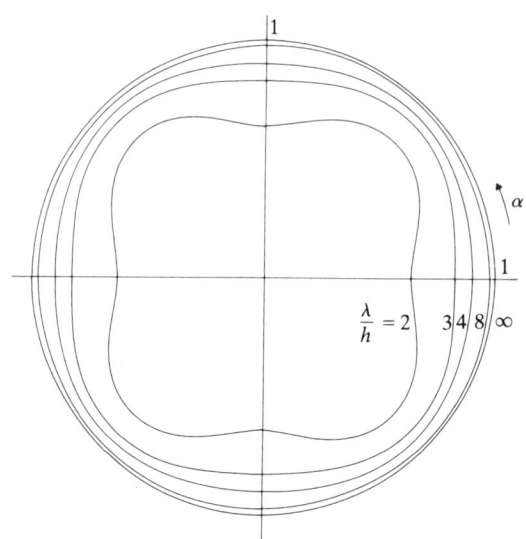

FIG. 10.7. *Polar diagram showing the normalized phase velocity c^*/c as a function of wavelength λ and direction α for the semi-discretization (10.23) of the wave equation. This figure is identical to Fig. 10.2 but with h replaced by $h/2$ (i.e., $\lambda/h = 2$ here is identical to $\lambda/h = 4$ in Fig. 10.2).*

EQUATIONS IN TWO DIMENSIONS: ANISOTROPY

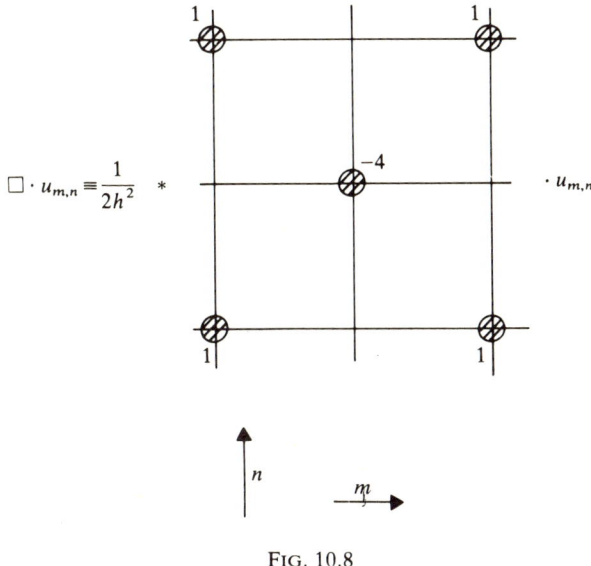

FIG. 10.8

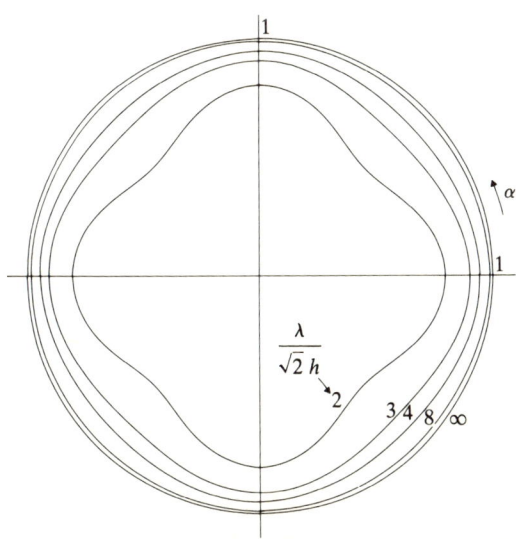

FIG. 10.9. *Polar diagram showing the normalized phase velocity c^*/c as a function of wavelength λ and direction α for the semi-discretization (10.34) of the wave equation. Note the 45° rotation in the directions of preferential propagation as compared with Fig. 10.7. Note also the change by $\sqrt{2}$ in the wavelength.*

The corresponding numerical velocity $c^*(\omega, \alpha)$ is now given by[2]

(10.36) $$c^*(\omega, \alpha) = c\frac{\sqrt{2}}{\omega h}(1 - \cos(\omega_x h)\cos(\omega_y h))^{1/2}.$$

As shown in Fig. 10.9, the directions of preferential velocity for sinusoidal waves are now the x- and y-axes. (They were the $\pm 45°$ directions in the preceding case (see Fig. 10.7).)

10.6. 9-point semi-discretizations of the wave equation. General explicit 9-point semi-discretizations of the wave equation can be obtained by a linear combination of the two preceding 5-point formulae (10.23) and (10.34). That is,

(10.37) $$\frac{d^2 u_{m,n}}{dt^2} = c^2(\beta\square\cdot + (1-\beta)\diamond\cdot)u_{m,n}$$

(where β is a parameter chosen in $[0, 1]$) represents a family of 9-point semi-discretizations.

The corresponding phase velocity is illustrated in Fig. 10.10 for several values of β.

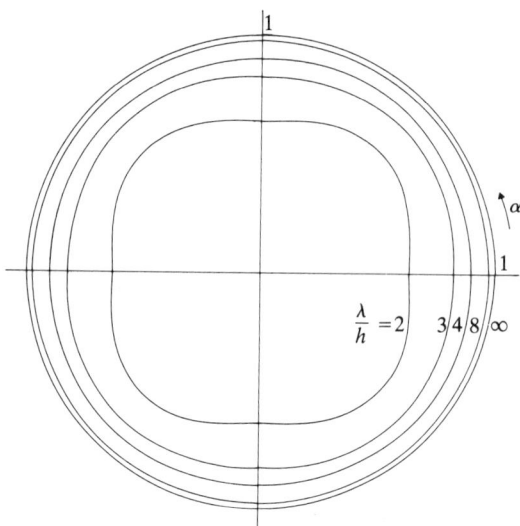

FIG. 10.10a. *Polar diagram showing the normalized phase velocity c^*/c as a function of wavelength λ and direction α for the semi-discretization (10.37) of the wave equation. Here $\beta = .25$.*

---

[2] Replace $E_x \cdot$ by $e^{i\omega_x h}$ and $E_y \cdot$ by $e^{i\omega_y h}$ in (10.35).

EQUATIONS IN TWO DIMENSIONS: ANISOTROPY

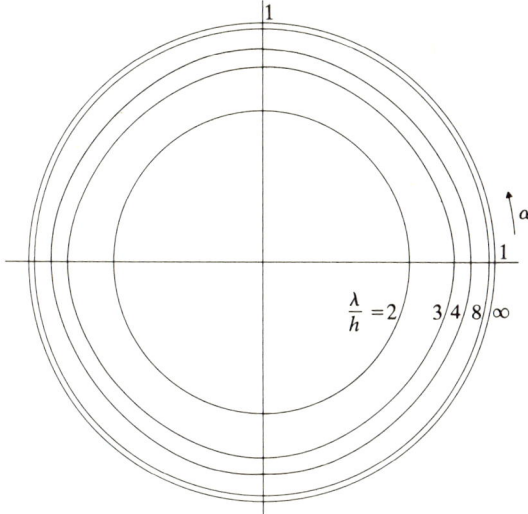

FIG. 10.10b. *Same as Fig.* 10.10a *except that* $\beta = .5$.

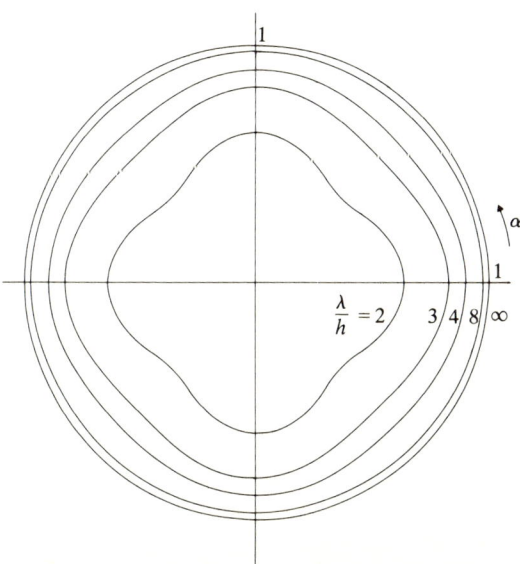

FIG. 10.10c. *Same as Fig.* 10.10a *except that* $\beta = .75$.

Bibliography

M. B. ABBOTT (1976), *Computational hydraulics: A short pathology*, J. Hydraulic Res., 14, pp. 271–285.

J. LE R. D'ALEMBERT (1747a), *Recherches sur la courbe que forme une corde tendue mise en vibration*, Mem. Acad. Sci. Berlin, 3 (1747: publ. 1749), pp. 214–219.

——— (1747b), *Suite des recherches sur la courbe que forme une corde tendue mise en vibration*, Mem. Acad. Sci. Berlin, 3 (1747: publ. 1749), pp. 220–253.

——— (1761), *Recherches sur les vibrations des cordes sonores*, Pamphlets, 1–73.

G. A. BAKER, V. A. DOUGALIS AND S. M. SERBIN (1981), *An approximation theorem for second order evolution equations*, Numer. Math., to appear.

P. BENYON (1968), *A review of numerical methods for digital simulation*, Simulation, 11.

D. BERNOULLI (1753a), *Reflexions et eclaircissemens sur les nouvelles vibrations des cordes...*, Mem. Acad. Sci. Berlin, 9 (1753: publ. 1755), pp. 147–172.

——— (1753b), *Sur le melange de plusieurs espèces de vibrations simples isochrones,...*, Mem. Acad. Sci. Berlin, 8 (1753: publ. 1755), pp. 173–195.

J. BERNOULLI (1727), *Theoremata selecta pro conservatione virium vivarum...*, Comm. Acad. Sci. Petrop., 2 (1727: publ. 1727), pp. 200–207.

——— (1728), *Meditationes de chordis vibrantibus.*, Comm. Acad. Sci. Petrop., 3 (1728: publ. 1732), pp. 13–28.

G. BIRKHOFF AND V. A. DOUGALIS (1975), *Numerical solution of hydrodynamic problems*, in Advances in Computer Methods for Partial Differential Equations, R. Vichnevetsky, ed., Publ. AICA, Rutgers University, New Brunswick, NJ.

G. BIRKHOFF AND S. GULATI (1974), *Optimal few-point discretizations of linear source problems*, SIAM J. Numer. Anal., 11, pp. 700–728.

G. D. BOOLE (1860), *A Treatise on the Calculus of Finite Differences*, Constable & Co., Ltd., London. Second edition reprinted by Dover, New York.

E. O. BRIGHAM (1974), *The Fast Fourier Transform*, Prentice-Hall, Englewood Cliffs, NJ.

L. BRILLOUIN (1946), *Wave Propagation in Periodic Structures*, McGraw Hill, New York.

——— (1960), *Wave Propagation and Group Velocity*, Academic Press, New York.

G. BROWNING, H.-O. KREISS AND J. OLIGER (1973), *Mesh refinement*, Math. Comp., Vol. 27, p. 29.

B. M. BUDAK (1961), *The method of straight lines for certain boundary value problems of the parabolic type*, Zh. Vychisl. Mat. i. Mat. Fix., (1961) pp. 1105–1112; U.S.S.R. Computational Math. Math. Phys., 1 (1962–63), pp. 1284–1291.

M. B. CARVER AND H. W. HINDS (1978), *The method of lines and the advection equation*, Simulation, 31, pp. 59–69.

R. C. Y. CHIN, G. W. HEDSTROM AND K. E. KARLSSON (1979), *A simplified Galerkin method for hyperbolic equations*, Math. Comput., 33, pp. 647–658.

R. V. CHURCHILL (1941), *Fourier Series and Boundary Value Problems*, McGraw-Hill, New York (revised edition 1963).

J. W. COOLEY AND J. W. TUKEY (1965), *An algorithm for the machine calculation of complex Fourier series*, Math. Comput., 19, pp. 297–301.

R. COURANT, K. FRIEDRICHS AND H. LEWY (1928), *Uber die partiellen Differenzengleichungen der mathematischen Physik*, Math. Ann., 100, pp. 32–74. (A translation by Phyllis Fox has been multilithed under the title "On the partial difference equations of mathematical physics," Report NYU-7689, Institute of Mathematical Sciences, New York University, 1956.

R. COURANT, E. ISAACSON AND M. REES (1952), *On the solution of nonlinear hyperbolic differential equations by finite differences*, Comm. Pure Appl. Math., 5, pp. 243–255.

V. A. DOUGALIS AND G. BIRKHOFF (1975), *A comparison of numerical methods for solving wave equations*, in First International Congress on Numerical Ship Hydrodynamics, J. W. Schot and N. Salvesen, eds., U.S. Naval Ship R & D Center (D. W. Taylor Model Basin).

V. A. DOUGALIS AND S. M. SERBIN (1981), *On the efficiency of some fully discrete Galerkin methods for second order hyperbolic equations*, Computers and Math. with Appl., to appear.

V. N. FADDEEVA (1949), *The method of straight lines in the application to certain boundary problems*, trans. into English from Trudy. Mat. Inst. Steklov., 28, pp. 73–103.

G. FIX AND G. STRANG (1969), *Fourier analysis of the finite element method in Ritz–Galerkin theory*, Stud. in Appl. Math., 48, pp. 265–273.

B. FORNBERG (1975), *On a Fourier method for the integration of hyperbolic equations*, SIAM J. Numer. Anal., 12, pp. 509–528.

J. FOURIER (1807), *Théorie de le propagation de le chaleur dans les solides*, Monograph presented to the Institut de France.

J. E. FROMM (1968), *A method for reducing dispersion in convective difference schemes*, J. Comput. Phys., 3, pp. 176–189.

M. C. GILLILAND (1966), *A spectral stability analysis of finite-difference operations*, IEEE Trans. Electronic Computers, EC-15, pp. 849–854.

D. GOTTLIEB AND S. A. ORSZAG (1977), *Numerical Analysis of Spectral Methods*, CBMS Regional Conference Series, 26, Society for Industrial and Applied Mathematics, Philadelphia.

A. R. GOURLAY AND J. LL. MORRIS (1968), *Finite difference methods for hyperbolic systems*, Math. Comp., 22, pp. 28–39.

—— (1972), *Hopscotch difference methods for non linear hyperbolic systems*, IBM J. Res. Development, pp. 349–353.

I. GRATTAN-GUINNESS (1970), *The Development of the Foundations of Mathematical Analysis from Euler to Riemann*, MIT Press, Cambridge, MA.

H. J. GRAY (1954), *Numerical methods in digital real time simulation*, Quart. Appl. Math., XII, pp. 133–140.

W. G. GRAY AND G. F. PINDER (1976), *An analysis of the numerical solution of the transport equation*, Water Resources Res., 12.

W. G. GRAY (1980), *Do finite element models simulate surface flow?*, in Finite Elements in Water Resources, S. Y. Wang et al., eds., School of Engineering, Univ. of Mississippi, University, MS.

B. GUSTAFSSON (1975), *The convergence rate for difference approximations to mixed initial boundary value problems*, Math. Comp., 29, pp. 396–406.

R. W. HAMMING (1977), *Digital Filters*, Prentice-Hall, Englewood Cliffs, NJ.

J. S. HICKS AND J. WEI (1967), *Numerical solution of parabolic partial differential equations with two-point boundary conditions by use of the method of lines*, J. Assoc. Comput. Mach., 14, pp. 549–562.

J. M. HYMAN (1979), *A method of lines approach to the numerical solution of conservation laws*, in Advances In Computer Methods For Partial Differential Equations, R. Vichnevetsky, ed., Pub. IMACS, New Brunswick, NJ, pp. 313–321.

L. B. W. JOLLEY (1961), *Summation of Series*, second revised edition, Dover, New York (first edition 1925, Chapman and Hall).

H.-O. KREISS (1964), *On difference approximations of the dissipative type for hyperbolic differential equations*, Comm. Pure Appl. Math., 17.

H.-O. KREISS AND J. OLIGER (1972), *Comparison of accurate methods for the integration of hyperbolic equations*, TELLUS XXIV, pp. 119–215.

────── (1973), *Methods for the Approximate solution of time dependent problems*, World Meteorological Organization/International Council of Scientific Unions, Geneva.

J. L. LAGRANGE (1759), *Recherches sur la nature de le propagation du son*, Miscell. Taurin, 1, also in "Works," Vol. 1, pp. 39–148.

C. LANCZOS (1956), *Applied Analysis*, Prentice-Hall, Englewood Cliffs, N.J.

P. D. LAX (1957), *Hyperbolic systems of conservation laws, II*, Comm. Pure Appl. Math., 10, pp. 537–566.

────── (1973), *Hyperbolic Systems of Conservation Laws and the Mathematical Theory of Shock Waves*, CBMS Regional Conference Series in Applied Mathematics, 11, Society for Industrial and Applied Mathematics, Philadelphia.

P. D. LAX AND B. WENDROFF (1960), *Systems of conservation laws*, Comm. Pure Appl. Math., 13, p. 217–237.

────── (1962), *On the stability of difference schemes*, Comm. Pure Appl. Math., 15, pp. 363–371.

P. D. LAX AND R. D. RICHTMYER (1956), *Survey of the stability of linear finite difference equations*, Comm. Pure Appl. Math., 9, pp. 267–293.

L. LAPIDUS AND J. H. SEINFELD (1971), *Numerical Solution of Ordinary Differential Equations*, Academic Press, New York.

B. LIU (1975), *Digital Filters and the Fast Fourier Transform*, collection of reprints of papers distributed by Halstead Press, division of John Wiley, New York.

A. M. LOEB AND W. E. SCHIESSER (1974), *Stiffness and accuracy in the method of lines integration of partial differential equations*, in Stiff Differential Systems, R. A. Willoughby, ed., Plenum, New York.

N. K. MADSEN AND R. F. SINCOVEC (1974), *The numerical method of lines for the solution of nonlinear partial differential equations*, in Computational Methods in Nonlinear Mechanics, J. T. Oden et al., eds., Texas Inst. for Computational Mechanics, Austin, TX.

W. L. MIRANKER (1971), *Difference schemes with best possible truncation error*, Numer. Math., 17, pp. 124–142.

G. MORETTI AND M. D. SALAS (1970), *Numerical analysis of viscous one-dimensional flows*, J. Comp. Phys., 5, pp. 487–506.

B. NOBLE (1969), *Applied Linear Algebra*, Prentice-Hall, Englewood Cliffs, NJ.

G. O'BRIEN, M. A. HYMAN AND S. KAPLAN (1950), *A study of the numerical solution of partial differential equations*, J. Math. and Phys., 29, pp. 223–251.

S. A. ORSZAG (1971), *Numerical simulation of incompressible flow within simple boundaries: Accuracy*, J. Fluid Mech., 49, pp. 75 ff.

A. PAPOULIS (1962), *The Fourier Integral and Its Applications*, McGraw-Hill, New York.

B. PARLETT (1966), *Accuracy and dissipation in difference schemes*, Comm. Pure and Appl. Math., 19, III.

LORD RAYLEIGH, 3rd BARON (J. W. Strutt) (1894), *The Theory of Sound*, Macmillan, London; reprinted by Dover, New York, 1945.

R. D. RICHTMYER (1962), *A survey of difference methods for non-steady fluid dynamics*, NCAR Technical Note 63-2, Nat'l. Center for Atmos. Res., Boulder, CO.

R. D. RICHTMYER AND K. W. MORTON (1967), *Difference Methods For Initial-Value Problems*, Interscience, New York.

P. J. ROACHE (1972), *Computational Fluid Dynamics*, Hermosa, Albuquerque, NM.

K. V. ROBERTS AND N. O. WEISS (1966), *Convective difference schemes*, Math. Comput., 20, pp. 272-299.

H. H. ROSENBROCK (1963), *Some general implicit processes for the numerical solution of differential equations*, Comput. J., 5, pp. 329-330.

E. L. RUBIN AND S. Z. BURSTEIN (1967), *Difference methods for the inviscid and viscous equations of a compressible gas*, J. Comput. Phys., 2, pp. 178-196.

I. V. SCHOENBERG (1946), *Contributions to the problem of approximation of equidistant data by analytic functions*, Quart. Appl. Math., 4; Part A, pp. 45-99; Part B: pp. 112-141.

I. J. SCHOENBERG (1973), *Cardinal Spline Interpolation*, CBMS Regional Conference Series in Applied Mathematics 12, Society for Industrial and Applied Mathematics, Philadelphia.

C. E. SHANNON AND W. WEAVER (1949), *The Mathematical Theory of Communication*, University of Illinois Press, Urbana.

G. STRANG (1962), *Trigonometric polynomials and difference methods of maximum accuracy*, J. Math. and Phys., 41, pp. 147-354.

——— (1971), *The finite element method and approximation theory*, in Numerical Solution of Partial Differential Equations, B. Hubbard, ed., Academic Press, New York, pp. 547-583.

B. SWARTZ (1975), *Comparing certain classes of difference and finite element methods for a hyperbolic problem*, in Advances in Computer Methods for Partial Differential Equations, R. Vichnevetsky, ed., Publ. AICA, Rutgers Univ., New Brunswick, NJ.

B. SWARTZ AND B. WENDROFF (1974a), *The relative efficiency of finite difference and finite element methods I: Hyperbolic problems and splines*, SIAM J. Numer. Anal., 11, pp. 979-993.

——— (1974b), *The relation between the Galerkin and collocation methods using smooth splines*, SIAM J. Numer. Anal., 11, pp. 994-996.

V. THOMÉE (1969), *Stability theory for partial difference operators*, SIAM Rev., 11, pp. 152-195.

——— (1973), *Spline-Galerkin methods for initial-value problems with constant coefficients*, Proc. Conference on the Numerical Solution of Differential Equations, Lecture Notes in Mathematics, 363, Springer-Verlag, New York.

L. TREFETHEN (1982), *Group velocity in finite difference schemes*, SIAM Rev., 24, pp. 113-136.

R. S. VARGA (1971), *Functional analysis and approximation theory in numerical analysis*, CBMS Regional Conference Series in Applied Mathematics 3, Society for Industrial and Applied Mathematics, Philadelphia.

R. VICHNEVETSKY (1971), *Numerical stability of methods of lines for the beam vibrations equation*, Report TR 71-18, Electronic Associates, Inc., West Long Branch, NJ.

——— (1972a), *Stability charts of methods of lines for partial differential equations*, Proc. Sixth Annual Princeton Conference on Information Science and Systems, Princeton Univ., Princeton, NJ.

——— (1972b), *The numerical treatment of boundary conditions in hyperbolic systems*, NAM-41, Dept. Computer Science, Rutgers Univ., New Brunswick, NJ.

——— (1973), *Physical criteria in computer methods for partial differential equations*, Proc. 7th AICA International Congr., Prague, reprinted in Proc. AICA, XVI, January, 1974, Brussels.

——— (1974), *Stability regions of two new families of explicit marching methods for partial differential equations*, NAM-149, Dept. Computer Science, Rutgers Univ., New Brunswick, NJ.

—— (1979), *Stability charts in the numerical approximation of partial differential equations: a review*, Math. Comput. Simulation, 21, pp. 170–177.

—— (1981a), *Energy and group velocity in semi-discretizations of hyperbolic equations*, Math. Comput. Simulation, 23, pp. 333–343.

—— (1981b), *Propagation through numerical mesh refinement for hyperbolic equations*, Math. Comput. Simulation, 23, pp. 344–353.

—— (1981c), *Computer Methods for Partial Differential Equations*, Vol. 1, Prentice-Hall, Englewood Cliffs, N.J.

R. VICHNEVETSKY AND A. W. TOMALESKY (1971), *Spurious error waves in numerical approximations of hyperbolic equations*, Proc. 5th Princeton Conference on Information Science and Systems.

R. VICHNEVETSKY AND B. PEIFFER (1975), *Error waves in finite element and finite difference methods for hyperbolic equations*, in Advances in Computer Methods for Partial Differential Equations, R. Vichnevetsky, ed., Publ. AICA, Rutgers Univ., New Brunswick, NJ.

R. VICHNEVETSKY AND F. DE SCHUTTER (1975), *A frequency analysis of finite element methods for initial value problems*, in Advances in Computer Methods for Partial Differential Equations, R. Vichnevetsky, ed., Publ. AICA, Rutgers Univ., New Brunswick, NJ.

R. VICHNEVETSKY AND S. SHYAM (1972), *Frequency analysis of methods of lines for hyperbolic systems*, NAM-50, Dept. Computer Science, Rutgers Univ., New Brunswick, NJ.

A. C. VLIEGENTHART (1969), *Dissipative difference schemes for shallow water equations*, J. Engrg. Math., 3, pp. 81–94.

J. VON NEUMANN AND R. D. RICHTMYER (1950), *A method for the numerical calculation of hydrodynamic shocks*, J. Appl. Phys., 21, pp. 232–237.

P. WESSELING (1972), *Accuracy of the third-order predictor corrector difference schemes for hyperbolic problems*, AIAA J., 10, pp. 948–949.

—— (1973), *On the construction of accurate difference schemes for hyperbolic partial differential equations*, J. Engrg. Math., 7, pp. 19–31.

E. T. WHITTAKER (1915), *On the functions which are represented by the expansions of interpolation theory*, Proc. Roy. Soc. Edinburgh, 35, pp. 181–194.

J. M. WHITTAKER (1929), *The Fourier theory of the cardinal function*, Proc. Edinburgh Math. Soc., 49, pp. 169–176.

Index

Advection equation, 2, 19, 80, 87, 94
    in two dimensions, 115ff
D'Alembert, J. le Rond, 1
Aliasing, 17
Amplification factor, 51, 58
Amplitude decay, 94
Amplitude error, 20, 59
Anisotropy, 115ff
Asymptotic approximation, 107, 108

Backward wave, 80-82
Baker, G. A., 61
Band-limited function, 16, 103, 104, 111
Band of frequencies, 88
Basis functions:
    linear, 6
    B-splines, 8
    trigonometric, 110
Bernoulli, John, 1
Birkhoff, Garrett, 33fn, 34fn, 124
B-splines, 8ff, 40
    Galerkin method, 8, 40ff
    collocation method, 42, 43, 47
    approximation of wave equation, 46
Boundary, 77, 85, 93, 94, 98, 100
    downstream, 97, 99, 101
    2-point and 3-point formulae, 101
Boundary conditions, 98, 109
    periodic, 109, 112
    sinusoidal, 93
Box method, 21, 59, 60, 62
Brillouin, L., 75
Browning, G., 94fn

Cardinal function, 13, 37
Characteristic equation, 86, 87, 92, 93
Characteristic form, 80, 83
Characteristic lines, 3
Characteristic roots, 86, 87, 93
Characteristic velocity, 81, 96
Chin, R. C. Y., 41
Collocation method:
    with B-splines, 42, 43, 47
    with Fourier series, 110, 111, 112, 114
Conservation, 88
    energy, 95
Conservation law, 29
Conservative approximation, 48, 107, 120
Consistent approximation, 80, 83, 87, 96
Consistent approximation to wave equation, 87
Constant amplitude, 88
Constant phase lines, 118
Convergence rates, 99, 107
Cosine method, 61
Courant number, 32, 54, 59, 60, 82, 124
Courant–Friedrichs–Lewy condition, 55
Crank–Nicolson method, 52, 59, 60
Cut-off frequency, 88, 92, 94

Directional vector, 116
Diffusion, spurious, 63
Discontinuity, 85, 97, 104
Discrete Fourier transform, 12
Dispersion, 24
Dissipation, 24
Dougalis, V. A., 34fn, 49, 61, 62, 124

INDEX

Eigenvalue, 112
Energy, 94
  conservation, 95
  propagation, 77, 97
  separation, 95, 97
Envelope, 75-77, 92
Equation residual, 5, 111
Error waves, 81
Euler's method, 11, 52, 58

Faddeeva, V. N., 5
Fast Fourier transform, 71, 113, 114
Fast Fourier transform filtering, 71
  frequency response, 73
Filtering, 69-74
Finite difference approximation, 123, 126, 129
  higher order, 35
  general, 9-point, 119-121, 129
  3-point, central, 77, 78, 80, 97
  upwind, 98, 102
  in time, 124
  general, 27, 28
Finite elements, 5ff, 26ff, 79
  bilinear, 122, 123
  linear, 6, 26
Finite element method, 5, 79
Fluid dynamics, 85, 112
Fornberg, B., 74, 112
Forward wave, 80-82
Fourier coefficients, 110, 111, 113
Fourier collocation method, 112, 114
Fourier components, 77, 88, 92
  discrete, 111
Fourier, J., 1
Fourier series, 76, 109
  truncated, 110, 112, 113
Fourier transform, 12, 23, 85, 86, 95, 98, 100, 103, 110
  of error, 100
  discrete, 95, 103, 104, 106, 108, 113
  in time, 85ff
Fully discrete approximations, 51
Fundamental solution, 86, 91, 92, 95, 97

Galerkin's method, 5ff, 111, 122, 123
  accuracy analysis, 26
General explicit time marching method (Vichnevetsky's method), 57, 58
General implicit time marching method, 56, 64

Global error, 23, 103ff
Group velocity, 75ff, 76(def), 94-96, 99
Gulati, S., 33fn
Gustafsson, B., 100

Hamming, R., 69
Heat equation, 1
Hedstrom, G. W., 41
Hicks, J. S., 5
Higher order semi-discretizations:
  explicit, 35, 37
  implicit, 40ff
Hyman, J. M., 5
Hyman, M. A., 54fn
Hyperbolic equation, 77, 85, 97, 108, 109, 111, 114
Hyperbolic equation in two dimensions, 115
Hyperbolic system, characteristic form, 2

Implicit semi-discretization in two dimensions, 120
Isotropic property of wave equation, 124, 125

Kaplan, S., 54fn
Karlsson, K. E., 41
Kreiss, H.-O., 112, 74, 17fn

$\mathscr{L}_2$-norm, 12, 13, 103ff
  of error, 104, 105, 107
  discrete, 103, 107
Lagrange, J. L., 1
Lanczos, C., 69
Laplacian, 123
Lax, P., 29fn
Lax–Wendroff method, 66
Leapfrog method, 12, 52, 58, 60, 81, 112, 113

Madsen, N. K., 5
Mesh refinement, 85
Method of lines, 5
Miranker, W., 108

Nodal value, 112, 113
Numerical damping, 65
Numerical experiments, 82-83, 106-107

O'Brien, G., 54fn
Oliger, J., 17fn, 74, 112, 94fn

Operator notation, 5
Order of accuracy, 25, 107
Orszag, S., 74, 112
Orthogonal functions, 15
Orthogonality property:
    of trigonometric functions, 100
    of cardinal function, 15
Oscillations, spurious, 68, 75, 77, 81, 83
Oscillatory part of numerical solution, 81, 83

Padé approximants, 53
Parseval's equality, 12, 15, 95, 103, 104
Phase lines, 116
Phase velocity, 75-78, 87-89, 92, 96, 105
    error, 21, 25
    directional dependence, 119
    in two dimensions, 117, 118, 120, 123, 125-128
Poisson's sum formula, 17
Polar diagram of phase velocity, 118, 120-123, 126-129
Propagation, 87
    preferential directions, 118, 127
    energy, 77
Pseudo-spectral method, 112

Rayleigh, Lord, (J. W. Strutt), 1
Recurrence equation, 86
Reflection ratio, 98, 100
Reflected solution, 91, 100-102
Reflection, 91, 97-99
    at a boundary, 97
    spurious, 97, 98
Richtmyer, R. D., 54fn
Roache, P., 94
Roberts, K. V., 22
Rosenbrock, H. H., 52
Runge–Kutta methods, 52, 57, 58

Sampling, 16, 104, 107
Sampling frequency, 17
Sampling of initial data, 16
Schoenberg, I. J., 8
Separation principle, 95
Serbin, S. M., 49, 61, 62
Shannon, C. E., 16
Semi-discretization, 4
    implicit, 7
Simpson's rule, 6

Sincovec, R. F., 5
Sinusoidal solutions, 103
Sinusoidal trial solutions, 19ff
Sinusoidal solutions in two dimensions, 116, 117, 124
Smooth part of numerical solution, 81, 83, 96
Space shift operation, 5, 86
Spatial frequency, 88, 91
Spatial operators in two dimensions, 119
Spectral methods, 109ff
Spectral function of an operator, 20, 112
Spectrum of an operator, 20
Spurious dispersion, 24
    of B-spline semi-discretizations,
    of the wave equation, 48
Spurious oscillations, 68, 75, 77, 81, 83
Spurious reflection, 97, 98
Spurious solutions, 77, 81, 97
Square grid, 116
Stability, 56, 112
Stability regions, 54-57
Stencil, 126
Stiffening (of propagating medium), 34
Störmer–Numerov formula, 33, 62
Strang, G., 43
Superconvergence, 45
Support of basis functions, 92
Swartz, B., 41, 43, 45, 46
Symbol of an operator, 20

Taylor series, 76, 100
Thomée, V., 20fn, 43
Time evolution, 81
Time frequency, 30, 85
Time-Fourier transform, 85ff
Time marching, 11
Toeplitz operators, 7, 40, 10
Tomalesky, A. W., 77
Trigonometric approximation, 109, 110, 112
Truncation error, 24-26, 33, 100, 101, 104

Upwind difference approximation, 64, 98, 102

Velocity, 82, 83, 110
Velocity vector, 115, 116
Vichnevetsky, R., 54fn, 57, 74, 77
Von Neumann, J., 54

Wave equation, 1, 31, 46, 66, 80, 87, 97, 124
   in two dimensions, 122, 128, 129
   isotropic property, 124, 125
Wave packet, 75, 76, 81, 77
   energy, 77
Wave propagation, 124
Wavelength, 75-79, 82, 87-89, 92, 93, 96, 116, 118, 119, 126, 127
Wei, J., 5
Weighted residuals (method of), 27, 79
Weiss, N. O., 22
Wendroff, B., 29fn, 41, 43, 45, 46
Wesseling, P., 108
Whittaker, E. T., 14
Whittaker's cardinal function, 13

RAYMOND H. FOGLER LIBRARY
DATE DUE

BOOKS ARE SUBJECT TO
RECALL AFTER TWO WEEKS

OCT 1 9 1987